AF768371

THE BIOLOGY OF THE SCHIZOPHRENIC PROCESS

ADVANCES IN BEHAVIORAL BIOLOGY

Editorial Board:

Jan Bures	*Institute of Physiology, Prague, Czechoslovakia*
Irwin Kopin	*National Institute of Mental Health, Bethesda, Maryland*
Bruce McEwen	*Rockefeller University, New York, New York*
James McGaugh	*University of California, Irvine, California*
Karl Pribram	*Stanford University School of Medicine, Stanford, California*
Jay Rosenblatt	*Rutgers University, Newark, New Jersey*
Lawrence Weiskrantz	*University of Oxford, Oxford, England*

Recent Volumes in this Series

Volume 8 • DRUGS AND THE DEVELOPING BRAIN
Edited by Antonia Vernadakis and Norman Weiner • 1974

Volume 9 • PERSPECTIVES IN PRIMATE BIOLOGY
Edited by A. B. Chiarelli • 1974

Volume 10 • NEUROHUMORAL CODING OF BRAIN FUNCTION
Edited by R. D. Myers and René Raúl Drucker-Colín • 1974

Volume 11 • REPRODUCTIVE BEHAVIOR
Edited by William Montagna and William A. Sadler • 1974

Volume 12 • THE NEUROPSYCHOLOGY OF AGGRESSION
Edited by Richard E. Whalen • 1974

Volume 13 • ANEURAL ORGANISMS IN NEUROBIOLOGY
Edited by Edward M. Eisenstein • 1975

Volume 14 • NUTRITION AND MENTAL FUNCTIONS
Edited by George Serban • 1975

Volume 15 • SENSORY PHYSIOLOGY AND BEHAVIOR
Edited by Rachel Galun, Peter Hillman, Itzhak Parnas, and Robert Werman • 1975

Volume 16 • NEUROBIOLOGY OF AGING
Edited by J. M. Ordy and K. R. Brizzee • 1975

Volume 17 • ENVIRONMENTS AS THERAPY FOR BRAIN DYSFUNCTION
Edited by Roger N. Walsh and William T. Greenough • 1976

Volume 18 • NEURAL CONTROL OF LOCOMOTION
Edited by Richard M. Herman, Sten Grillner, Paul S. G. Stein, and Douglas G. Stuart • 1976

Volume 19 • THE BIOLOGY OF THE SCHIZOPHRENIC PROCESS
Edited by Stewart Wolf and Beatrice Bishop Berle • 1976

A Continuation Order Plan is available for this series. A continuation order will bring delivery of each new volume immediately upon publication. Volumes are billed only upon actual shipment. For further information please contact the publisher.

THE BIOLOGY OF THE SCHIZOPHRENIC PROCESS

Edited by

Stewart Wolf

The University of Texas Medical Branch
Galveston, Texas

and

Beatrice Bishop Berle

Albert Einstein College of Medicine
New York, New York

PLENUM PRESS • NEW YORK AND LONDON

Library of Congress Cataloging in Publication Data

Totts Gap Colloquium on the Biology of Schizophrenia, 1975.
The biology of the schizophrenic process.

(Advances in behavioral biology; v. 19)

Includes index.
1. Schizophrenia — Congresses. 2. Mental illness — Physiological aspects — Congresses. 3. Mental illness — Genetic aspects — Congresses. I. Wolf, Stewart George, 1914- II. Berle, Beatrice Bishop III. Title. [DNLM: 1. Schizophrenia — Etiology — Congresses. 2. Schizophrenia — Therapy — Congresses. 3W TO721 1975b/[WM203 T721 1975b]
RC514.T64 1975 616.8'982'071 76-24828
ISBN-13: 978-1-4615-8995-2 e-ISBN-13: 978-1-4615-8993-8
DOI: 10.1007/978-1-4615-8993-8

Proceedings of the Totts Gap Colloquium on The Biology of Schizophrenia,
held in Totts Gap, Pennsylvania, May 29-31, 1975

Officers and Directors:

John G. Bruhn, Ph.D., *President and Director*
Beatrice B. Berle, M.D., *Vice-President and Director*
Stewart Wolf, M.D., *Secretary Treasurer and Director*
Mr. George Plush, *Director*

Board of Trustees:

Mark D. Altschule, M.D.
William B. Bean, M.D.
Robert M. Bird, M.D.
Andre Cournand, M.D.
William C. Gibson, M.D.
Miss Helen Goodell
Franz J. Ingelfinger, M.D.
Robert C. Page, M.D.
Edmund D. Pellegrino, M.D.
Eliot Stellar, Ph.D.
Mr. Oscar Swarth
Joseph M. White, M.D.

Advisory Council to Totts Gap Colloquia:

Mark D. Altschule, M.D.
Andre Cournand, M.D.
Martin M. Cummings, M.D.
Franz J. Ingelfinger, M.D.
Mr. Oscar Swarth
N. T. Werthessen, Ph.D.

© 1976 Plenum Press, New York
Softcover reprint of the hardcover 1st edition 1976
A Division of Plenum Publishing Corporation
227 West 17th Street, New York, N.Y. 10011

All rights reserved

No part of this book may be reproduced, stored in a retrieval system, or transmitted, in any form or by any means, electronic, mechanical, photocopying, microfilming, recording, or otherwise, without written permission from the Publisher

Photo by Thomas Wolf

PARTICIPANTS IN THE COLLOQUIUM

From left to right and from above to below: Snyder, Wyatt, Wolf, Berle, Race, Bliss, Cohen, Bolen, Bird, Gjessing, Werthessen, Altschule, Cancro, Bigelow, Lipton, Haber, Undesser, Goodell, Goldstein, Kety, Hughes, Carson, Rosenfeld, Boswell-Styles, Beyer, Swarth.

Photo by Thomas Wolf

Dr. Seymour Kety with student, Eric Undesser, and other students during luncheon break.

Preface

This volume contains the edited transcript of the second Totts Gap Colloquium, held at Totts Gap, Pennsylvania, May 29-31, 1975. The objective of the Colloquium was to bring into focus what is known of pathogenic mechanisms of schizophrenic manifestations. One hope was to differentiate among syndromes that share the features of schizophrenia, disorganization of thought and talk, inappropriateness of affect, circumstantiality, withdrawal from social interaction, paranoid ideation, hallucinations, delusions, catelepsy, catatonic excitement or mutism, etc. What is known of the neural and chemical processes capable of inducing such aberrations was discussed with an attempt to reconcile disparate findings and interpretations, to synthesize current evidence and to point up crucial areas of ignorance that need exploration. The program was designed to further the objectives of the Multidisciplinary Research Program in Mental Health, of the University of Texas Medical Branch and to supplement their efforts with a well focused look at the schizophrenic process. Accordingly, Dr. Allan Goldstein, Professor of Biochemistry and Acting Director of the Multidisciplinary Research Program in Mental Health; Dr. Bernard Haber, Associate Professor of Biochemistry and Member of the Marine Biomedical Institute; and Dr. Geraldine Cohen, Assistant Professor, Department of Human Biological Chemistry and Genetics, joined in discussion with the Colloquium participants who were as follows:

Dr. Mark D. Altschule, Clinical Professor, Harvard Medical School.
Dr. Llewellyn B. Bigelow, Director of Mental Health Research, St. Elizabeth's Hospital.
Dr. Eugene L. Bliss, Professor and Chairman of Psychiatry, University of Utah.
Dr. Robert Cancro, Professor of Psychiatry, University of Connecticut.
Dr. Leiv R. Gjessing, Chief, Central Laboratory, Dikemark Hospital, Solberg, Norway.
Dr. Seymour Kety, Professor of Psychiatry, Harvard University.

Dr. Morris Lipton, Professor and Chairman of Psychiatry, University of North Carolina.
Dr. Solomon H. Snyder, Professor of Pharmacology and Psychiatry, Johns Hopkins University.
Dr. Richard J. Wyatt, Chief, Laboratory of Clinical Psychopharmacology, St. Elizabeth's Hospital.
Dr. Stewart Wolf served as moderator.

Six medical and graduate students at the University of Texas Medical Branch attended as observers:

Ms. Carol B. Beyer (Medical Student)
Mr. Dan Bolin (Medical Student)
Ms. Tracy Hughes (Medical Student)
Harrol T. Hutchison, Ph.D. (Medical Student)
Mr. Bill Race (Medical Student)
Mr. Eric Undesser (Graduate Student)

Financial support for the Colloquium was provided by Geigy Pharmaceutical Company who also supplied the original transcription of the proceedings. Recording equipment was provided through the generosity of Mr. Oscar Swarth. Grateful acknowledgement is also made to Mrs. Joan Martin, Miss Colleen Hogan and Mrs. Hannah Norton who undertook the task of organizing edits and preparing the final manuscript.

The following Trustees of Totts Gap attended the Colloquium:

Dr. Mark D. Altschule (also a participant)
Dr. Beatrice B. Berle
Dr. Robert M. Bird
Miss Helen Goodell
Dr. E. Pellegrino
Dr. Eliot Stellar
Mr. Oscar Swarth

Mr. Albert Rosenfeld, Science Editor of the SATURDAY REVIEW and Consultant to the National Foundation, also attended. Mr. Robert Boswell-Styles and Mr. Vincent L. Carson attended from the Geigy Company.

S.W. and B.B.B.

Contents

Introduction

For three hundred years schizophrenia has provided medicine with one of its most tantalizing puzzles. Only recently have glimpses of understanding begun to penetrate the dark mystery of this common and devastating disorder. As summarized in this volume, there is now convincing evidence of genetic transmission of at least one major form of schizophrenia. Another shaft of light comes from studies of brain biochemistry, especially as reflected in pharmacologic data from therapeutic agents on the one hand and from pharmacologic simulators of schizophrenia syndromes on the other.

The role of life experience and environmental forces in the pathogenesis of schizophrenia remains elusive although the frequent failure of schizophrenia to appear in one of a pair of monozygotic twins suggests that psychosocial forces may be causative in a genetically predisposed individual. Other potential factors considered in this volume are assaults on the central nervous system including birth trauma, minimal brain damage, infections and other stresses. In addition immunologic mechanisms and even slow viruses are brought into consideration to explain the penetrance of a schizophrenic genetic trait.

The participants in this Colloquium, pictured on page engaged in an open, uninhibited and thoughtful analysis of available data from all of these quarters. Psychosocial pressures were considered as well as genetically and behaviorally induced disorders in animal models. Six medical and graduate students were included in the Colloquium as observers. An up to date, reasoned critique of the evidence, regarding the nature and pathogenesis of schizophrenia was achieved and some directions for future inquiry were identified.

CHAPTER I - Historical Perspective - Evolution of the Concept of Schizophrenia

Participants: Drs. Altschule, Bliss, Cancro, Kety, Snyder

DR. ALTSCHULE: My presentation will be in three parts: (1) preliminary remarks on classification; (2) a summary of the history of the schizophrenia concept; and (3) comments that will enlarge upon or illuminate some portions of this history.

Schizophrenia as a Syndrome

In the first place, it is quite clear that the history of the development of the schizophrenia concept is the history of the development of ideas about a group of symptoms. Although the 18th and 19th century writings on mental disease refer to a multitude of post-mortem studies there is no indication anywhere that the writers believed that specific brain lesions gave rise to specific symptoms. (In passing it should be noted that similarly there was no indication that the various symptoms or groups of symptoms that were noted differed clinically from each other as regards purported etiology). The accumulated clinical and pathological evidence of recent years can lead only to the conclusion that what we call schizophrenia is not a disease but only a clinical syndrome. (We shall dismiss as unintelligible the notion put forward by some particularly muddle-headed persons who hold that it is merely a <u>state of mind</u>, whatever that is.)

If we agree - as all the available evidence demands - that what is called schizophrenia is a syndrome, we must establish whether or not it is merely a figment of classification. We recognize that classification is, in the words of Lewis Carroll's writings on symbolic logic, a process whereby we imagine that certain things should be grouped together. This states correctly that the products of classification are the products of thinking. Does this mean that they have no reality - are purely imaginary? We shall leave to one side the Platonic notion that anything that has a name must exist -- in fact the only reality is the idea of a thing, but we should not dismiss consideration of how - and why - names develop. We must address ourselves to the problem of whether entities derived from classification can justify their existence. This does not involve questions about reality, and in fact any such questions would divert us from the purposes of our inquiry. Obviously when people see something, a name for it is soon forthcoming. This process is called the Eden Maneuver, because it describes what the Bible says happened when the animals paraded before Adam - the only person in authority, in fact the only person at all who was present. In medicine when we have a name for a disease, a name bestowed by some person of recognized authority, the attributes connected to that include: (1) the responsible specific etiological factors; (2) the processes involved in pathogenesis; (3) the physiology and chemistry responsible for its signs and symptoms; (4) and the signs and symptoms - including the societal consequences - of the disease. There may be disagreement about details but all physicians know what the disease is recognized to be.

The situation is different with respect to syndromes. We may recognize a number of widely different etiologic factors, or we may recognize none at all. We may recognize a common pathogenesis, or none at all. We may recognize a common disturbance of physiology or chemistry, or we may recognize none at all. But in every case we recognize an identity of signs and symptoms, and this is what determines the clinical validity of the syndrome.

Three centuries ago the clinical syndrome of fever had no proven etiology, no established pathogenesis, no recognized disturbances of physiology or chemistry, yet nobody will deny that it was a valid syndrome. It had signs and symptoms that clinicians agreed upon, even though they could not be measured objectively. Can we say the same thing about schizophrenia today? The answer is an unequivocal yes.

The Development of the Schizophrenia Concept

In the 17th century when the concept of schizophrenia was developing, there was nowhere in medicine any conviction concerning specific etiologies for any diseases. Hence the etiology of mental diseases was the same as for all others, i.e., psychological factors, environmental factors and humoral factors. All diseases were in effect no more than syndromes, and this applied also to mental disorders. The germ theory of disease, and some toxicological disorders, resulted in the development of the idea of specific etiologies, but this came much later.

Post-mortem studies in mental diseases revealed in many instances evidence of severe brain disease, i.e. syphilis, tumors, cysts, etc. was present. Hence if there was any idea that the mental diseases had an organic representation, this was still non-specific. There is nowhere any indication in the history of the development of the schizophrenia concept that any author believed it to be a specific disease in the modern sense.

The present discussion is a major portion of the section on thought disorders in a soon to be published work (4). This account of the development of the schizophrenia concept will make it clear that the concept, although rooted in ancient clinical observations, took form slowly, and reached maturity late. The manic-depressive psychosis was clearly described, in writing, eighteen centuries ago and a question naturally arises as to why this psychosis was recognized as an entity centuries before the other main psychiatric syndromes received clear recognition. This is easily answered: What is considered depression or elation may vary somewhat from time to time and place to place but in the main they are easily recognized when they exceed the limits of normal day to day variations in mood. Disordered thinking, manifested by delusions, on the other hand, is much harder to recognize, unless extreme. This disorder, in its milder forms, manifests itself only by deviations from the norm as regards occupation, ambition, manners, attitude toward education, personal behavior, etc., and as such may be considered on the one hand as personal idiosyncracy or on the other as an antisocial attitude. These diagnostic difficulties are especially likely to arise in a highly complex society where these behavior patterns may simply be manifestations of a considered unwillingness to respond to social stresses in accepted - usually stereotyped - ways. Hence simple observations of behavior often could not help to diagnose thinking disorders. It was not until the development of associationist psychology occurred in the 18th century that the interpretation of thinking disorders was given direction. This could not have happened in classical Greece, which gave us fully formed the concept of manic-depressive psychosis. For one thing, hallucinations were considered objective phenomena that happened to be seen or heard only by selected persons. The distorted thinking processes that today we call schizophrenia were not considered to constitute a separate disorder. In the Middle Ages the dominance of soul-psychology and of demonology, both of which had the advantage of being able to explain anything, prevented any rational approach to the study of psychiatric phenomena. When careful clinical observation later became the method of medicine, the more or less specific details of clinical syndromes came to be recognized.

Dementia

Dementia was among the most ancient of the diagnostic categories in psychiatry. It was distinguished from amentia in that it represented a loss of reasoning power, whereas in amentia this power never developed. It was first sub-divided according to the age at which it occurred, rather than in accordance with its clinical features, as Dufour noted (43).

The origin of the schizophrenia concept has its roots in dementia praecox. This statement is valid despite Sullivan's belief that the two are distinct disorders (177).

Thomas Willis (1683) pointed out that dementia may be a sequel of protracted melancholia (187). This perceptive observation was repeated, usually without attribution, by a host of 19th century authors.

Hartley's ESSAY ON MAN with its extended exposition of the phenomena of mental association, stimulated the evaluation of psychiatric disorders in relation to these phenomena (75). A number of 18th century psychiatric writers made comments along these lines. Crichton was evidently the first to refer to looseness of associations in dementia (37).

The essential psychologic disorder of what today is called schizophrenia was defined by Pinel in 1809 under the term démence or dementia (146). My translation of Pinel's word 'disparate' is unfitting or unsuitable, rather than unequal as is frequently given.

Haslam (1809) offered the first clear picture of a schizophrenic psychosis. He gave the disorder no name, but his description of the onset at puberty and the progressive deterioration indicate that he was in fact the first to describe what later came to be called hebephrenia. He wrote (76):

> " Connected with loss of memory, there is a form of insanity which occurs in young persons; and, as far as these cases have been the subject of my observation, they have been more frequently noticed in females. Those whom I have seen, have been distinguished by prompt capacity and lively disposition: and in general have become the favorites of parents and tutors, by their facility in acquiring knowledge, and by a prematurity of attainment. This disorder commences, about or shortly after, the period of menstruation, and in many instances has been unconnected with hereditary taint; as far as could be ascertained by minute enquiry. The attack is almost imperceptible; some months usually elapse before it becomes the subject of particular notice; and fond relatives are frequently deceived by the hope that it is only an abatement of excessive vivacity, conducing to a prudent reserve, and steadiness of character. A degree of apparent thoughtfulness and inactivity precede, together with a diminution of the ordinary curiosity, concerning that which is passing before them; and they therefore neglect those objects and pursuits which formerly proved sources of delight and instruction. The sensibility appears to be considerably blunted; they do not bear the same affection towards their parents and relations; they become unfeeling to kindness, and careless of reproof. To their companions they shew a cold civility, but take no interest whatever in

> their concerns. If they read a book, they are unable to give any account of its contents: sometimes, with steadfast eyes, they will dwell for an hour on one page, and then turn over a number in a few minutes. It is very difficult to persuade them to write, which most readily develops their state of mind: much time is consumed and little produced. The subject is repeatedly begun, but they seldom advance beyond a sentence or two: the orthography becomes puzzling, and by endeavouring to adjust the spelling the subject vanishes. As their apathy increases they are negligent of their dress, and inattentive to personal cleanliness. Frequently they seem to experience transient impulses of passion, but these have no source in sentiment; the tears, which trickle down at one time, are as unmeaning as the loud laugh which succeeds them; and it often happens that a momentary gust of anger, with its attendant invectives, ceases before the threat can be concluded. As the disorder increases, the urine and feces are passed without restraint, and from the indolence which accompanies it, they generally become corpulent. Thus in the interval between puberty and manhood, I have painfully witnessed this hopeless and degrading change, which in a short time has transformed the most promising and vigorous intellect into a slavering and bloated idiot."

The concept of dissociation, now regarded as a modern invention was actually advanced more than a century and a half ago. Benjamin Rush (157), that sulky, irascible genius, chose one of several psychological disorders attached to dementia by Pinel, and used dissociation as a synonym for Pinel's démence. Rush, being a Colonial, his contribution was not widely recognized. Actually no American psychiatrists received much attention until the meteoric but short-lived fame of James Beard changed the situation.

Rush's idea of dissociation led to the word schizophrenia being invented a century later, at a time when the suffix -phrenia was at the height of its popularity.

Jacquelin-Dubuisson in 1816 referred to dementia in what was by then the established manner but with more detail than given by his predecessors. The clinical distinction, as regards course and outcome, between acute and chronic dementia was noted. That dementia might be a sequel of some other psychoses was also stated. He did not, however, separate the psychologic features of some types, e.g. senile dementia, sharply enough (94).

Georget was even more emphatic than others in making dementia the terminal state of all incurable psychoses (62). He mentioned the non-specificity of dementia's precursors and sequels. He also stressed the hebetude of the patients and noted that it was not necessarily incurable. He observed that shock (e.g. fever) may have a curative effect.

The clinical description offered by Burrows in 1820 exemplifies the development of a clinical concept as more detailed observations are recorded. His debt to Pinel is obvious (17). He wrote:

> "Fatuity or demency, however, must not be confounded with imbecility or idiotism: it is materially different from either. The faculties of a person in demency, though perhaps alienated, are not always abolished; they are often only in abeyance; and may revert to a state of sanity, either spontaneously or from judicious treatment; or they may be only partially affected. But assuredly, if the case be misjudged, as it too often is, the patient may pass into perfect idiocy ...
>
> There is a marked difference between demency and imbecility or idiocy. The connate imbecile never had his mental faculties fully developed, and therefore never attains the ordinary standard of adult intelligence. True demency implies a previous possession and exercise of the mental faculties; but one or more of those faculties may, by accident, disease, or age, have become simply deteriorated. The conversation and manners of the latter preserve some traits of the character of the original man; those of the former always retain the impress of childhood. The connate idiot never possessed any intellectual endowment ... "

Conolly, as was his wont, described the early subtle changes in personality that might usher in a psychotic episode (33).

To Guislain in 1833, dementia, although a disorder of intellectual function, was basically a disorder of affective processes (73).

Pritchard's succinct statement in 1837, emphasized the weakness of associations in dementia (148).

Esquirol gave a detailed analysis of the effects of weakened associations and also emphasized the passivity (47).

Dementia Praecox first used

Early and mid-19th century authors described the syndrome of primary dementia and in this they followed Pinel, Haslam and their immediate followers. They distinguished primary dementia from secondary dementia. Many instances of the latter undoubtedly representing syphilis of the brain, the rest including examples of dementia following endogenous mania or melancholia. The authors who discussed primary dementia followed the example of Haslam; they paid little detailed attention to the earlier manifestations. Morel (1860) remedied this defect (128). He wrote: "An unfortunate father consulted me one day about the mental state of his 12 or 14 year old son, in whom a violent hatred for the originator of his life had suddenly replaced the most tender feelings. He was despondent at being the smallest in his class, despite the fact that he was always first in his compositions, and that without effort and almost without work. It was, so to speak, by intuition that he comprehended things, which he organized in his memory and his intelligence. He gradually lost his cheerfulness, became gloomy, taciturn, and showed a tendency toward solitude. His mother was psychotic and his grandmother eccentric to an extreme degree. I ordered the interruption of the child's studies and his confinement in an institution for hydrotherapy. A most happy change occurred in the bodily state of the child. He grew considerably but another phenomenon as disquieting as those I have mentioned came to dominate the situation. The young patient progressively forgot everything he had learned; his so brilliant intellectual faculties underwent in time a very distressing arrest. A kind of torpor akin to hebetude replaced the earlier activity, and when I saw him again I concluded that the fatal transition to the state of démence précoce was about to take place. This dreadful prognosis is ordinarily far from the minds of the parents and also of the physicians who care for these children. Such is, nevertheless, in many cases the sad termination of hereditary insanity. A sudden paralysis of the faculties, a démence précoce, indicate that the patient has reached the end of his intellectual life that he can control."

Morel used the descriptive words so casually, and not as a diagnostic term (the index of his book does not refer to démence précoce or any derivative of it), that they evidently made no impression on most other psychiatrists or even on himself.

Dickson treated "infantile dementia" as a separate category in 1874 but it is clear from his descriptions that he had grouped together the results of a variety of organic brain syndromes. Dickson's discussion (41) is noteworthy in several respects: (1) it mentioned what may have been autism; (2) it recognized that the term dementia needed clarification; and (3) it did not use the term dementia praecox.

Dementia Praecox as a diagnostic category

The first occurrence of the words démence précoce as a diagnostic term is in the title of Gauthier's 1883 Paris Thesis. His patients' symptoms included delusions, hallucinations, paranoia and catatonia. He nowhere mentions Morel (61).

Clouston brought the psychoses of adolescence to the fore in 1888 although he preferred the term secondary dementia and did not mention dementia praecox. Later Krafft-Ebing showed the same preference (110). He wrote:

> " I answer the question 'what is the typical form of insanity' by saying that the insanity of adolescence is the typical form, because it most frequently ends in typical secondary dementia, without any other function being affected but mentalization, and because in its course we have all the forms of psychosis represented ... Almost all pure cases of secondary dementia will be found to have originated in the developmental (pubescent and adolescent) insanities ... Undue and unphysiological means through a forcing-house mode of education during adolescence without regard to the heredity, capacity and weaknesses of the organism tend toward dementia. The constant changes in each generation of modern civilized life in the adaptation of the human organism to its environment and the special efforts thus rendered necessary by the struggle for existence tend towards dementia through the strain they put on the most delicate of all organized tissues."

Charpentier muddied the waters in 1891 when he used the term dementia praecox to mean any mental disorder in a young person, designating eleven groups (30).

A short time later in 1891 A. Pick of Prague published a paper on dementia praecox and referred to both Morel and Gauthier in it (144). Kraepelin referred to Pick as the originator of the concept.

However, Tuke's Dictionary (1892) does not include dementia praecox and other authorities continued to prefer secondary dementia.

Hebephrenia

The origin of the term hebephrenia is as follows. In the mid-19th century the use of phren as a prefix or suffix became established. For example, in 1852 Guislain suggested using phrenalgia for depression, phrenoplexia for catatonia, hyperphrenia for mania, ideophrenia for impulsive insanity, and aphrenia for dementia. A decade later Kahlbaum suggested dysphrenia for secondary insanity, and paraphrenia for age-related psychoses. The group was further divided into neophrenia for infantile brain dysfunctions, hebephrenia for pubertal psychoses, and presbyphrenia for senility. (Some decades later the trend was continued in schizophrenia.) Of Kahlbaum's list of terms only hebephrenia came to be widely accepted. This was largely owing to Hecker's writing on the subject.

Plato stated that if anything has a name it must exist as a real thing. It was therefore to be expected that one of Kahlbaum's pupils, Hecker, should write about hebephrenia as if it were really a distinct disorder and not a figment of classification (79).

Catatonia

Catatonia received its name late in the 19th century, but it had been recognized a century earlier. It received passing mention in clinical accounts that described its occasional occurrence in what was then called melancholia atonica. Guislain defined the syndrome in 1833 and gave it the name Ecstasy (72).

The writings of Gooch, a specialist in diseases of women (71) contain a remarkable early account of the entire course of a catatonic attack.

Griesinger's general description, more detailed as regards the mental state, followed that of Guislain (71).

Catatonia is sometimes called Kahlbaum's Syndrome. Although earlier authors had clearly referred to the physical manifestations of the disorder, he was the first to define its psychopathology. It is worthy of note that he emphasized its cyclic nature, the occurrence of manic, melancholic and paranoid symptoms during its course, and the terminal development of dementia in some cases. In this respect he must be considered the forerunner of Kraepelin who included catatonia as a syndrome of dementia praecox. Nevertheless Kahlbaum maintained that the unique basic process was negativism, both psychologic and somatic (98).

Kahlbaum, who believed that all mental disorders of childhood were best treated by means of special education, operated a school for disturbed children, some of whom had catatonia, wrote in 1884 (99):

> " It is well known that most mental diseases can appear in youth, especially dysthymia and vesania which strangely enough are still called melancholia and mania; further catatonia may occur in its simpler form of the so-called atonic melancholy as well as in the more complex form with manic and choreactic symptoms ..."

Spitzka wrote about catatonia shortly after Kahlbaum's publication. Spitzka followed Kahlbaum in general but with some different emphasis, and, in one respect, some disagreement (173); he did not believe that the use of diminutives was typical.

Paranoia

The word paranoia originally had a meaning different from that given it today. At first it meant nothing but "mentally distraught." According to Professor Renehan of Boston College, the word is found in several places in ancient Greek drama, i.e. in SEVEN AGAINST THEBES by Aeschylus (v. 756) presented in 467 B.C.; in ORESTES by Euripides (v. 824) presented in 408 B.C.; and in CLOUDS by Aristophanes (v. 1976) perhaps presented in 423 B.C.; or else in the revised version later. In all cases the word paranoia was used simply to indicate mental disorder. Hippocrates used it in a similar sense.

Vogel is known to have used the word paranoiae as the general heading for his group of mental diseases in his works before 1770, but by 1772 he had dropped it and replaced it by morbi mentis (182). Heinroth continued to use it as Vogel had, and other German clinicians did likewise. However, it came to have a new and highly specific meaning when Kahlbaum in 1878 suggested that paranoia be reserved for the primary form of systematized delusions to distinguish it from the form in which it occurred secondarily in other psychoses. The confused histories of both the word paranoia and of the syndromes to which it was or was not attached is described in detail by Seglas, who wrote: "There is perhaps no word in psychiatry that has a wider acceptance or a vaguer definition." (162)

The word paranoia was used as a formal diagnosis in the first edition of Krafft-Ebing's text (1879). Spitzka discussed the condition without using the name in his first edition, but used it in his second.

Although writers of Haslam's era and before had described delusions of persecution, Haslam's 1810 report was the first systematic and detailed account of a patient who seemed normal in all respects except those relating to his systematized delusions (77). When the patient was examined by two famous physicians, the immortal Clutterbuck and the outstanding Birkbeck, they declared him completely sane.

Laseque's paper (113) was a highly important contribution to the development of a psychiatric concept, a contribution that has been almost entirely unrecognized. He stated: (1) that delusions of persecution are common in all types of psychosis; (2) that they may be the primary manifestation of mental disease, unassociated with other symptoms, and (3) most important of all, that the basic psychologic process is an intense pre-occupation with explaining the delusions logically. The last feature has received only tangential recognition in today's comment that there is nobody so logical as a paranoid.

Laseque's concept of the nature of paranoia fits what is observable about the disorder and hence is more satisfactory than Freud's trivial explanation that paranoia is a manifestation of latent homosexuality.

Morel first named what we call today paranoia in 1852, calling it folie systematisée or manie systematisée. However, he was describing the primary syndrome and gave it the name delire de persecutions. Morel later adopted this term for the secondary variety also (128).

Ball described the manner in which the patient's delusions completely engulfed the individual, despite the outward lucidity and reasonableness manifested in other areas of mental life (9).

Kraepelin's Classification

Kraepelin needed a new term by which he hoped to unite the entire group of endogenous psychoses of the young and he chose dementia praecox (108). He considered the term particularly apt because of the high incidence of mental deterioration in the patients.

Kraepelin is currently derogated because he emphasized deterioration. However it is clear from his writings that what he considered deterioration was a remission of the active psychosis but involved a return to a diminished personality less effective than it had been before the onset of the illness. Today, especially in this country, we shun the word deterioration and may classify such a remission as a cure in order to improve the statistics.

It is apparent that by the end of the 19th century the schizophrenia concept, although as yet unnamed, had a solid existence. Kraepelin's choice of the term dementia praecox to designate it was probably ill-advised. The term was not justified historically; it implied, inaccurately, the invariable occurrence of deterioration, and it was interpreted as always indicating a development in childhood or youth (although Kraepelin used the word praecox merely to distinguish it from senile dementia). Nevertheless, his clinical descriptions were unsurpassed, and were matched for broadness, thoroughness, and perceptiveness only by those of von Krafft-Ebing (whose rising star unfortunately fell rapidly after he published a pornographic book, something that a "von", i.e. a man of noble birth, simply never did) (110).

Bleuler's Contribution - Schizophrenia

The question arises as to what Bleuler contributed to the problems of definition and nomenclature. It is clear that all the pieces from which he constructed his clinical concept were available and, at least in Europe, well known. Although Scotch tape had not yet been invented, he did very well without it. Bleuler defined the condition we are discussing here as one in which the most important basic psychologic process was dissociation. Although he added other "primary" phenomena, dissociation was the one that, in his view, gave the disorder its name, schizophrenia. This was entirely solipsistic, but inevitable since his mentor, Kahlbaum, had already made the suffix phrenia popular. This choice gave supremacy to this one manifestation, a manifestation given prominence by a considerable number of psychiatrists beginning with Pinel. It cannot be defended on any grounds except perhaps, divine inspiration. Nevertheless, he gave us a term that became widely accepted. The term has a certain attractiveness: it is in the historical tradition of the use of -phrenia established by one of psychiatry's greatest minds, Kahlbaum; although euphonious, it is operationally meaningless.

DR. KETY: What is the literal meaning of the Greek word 'phrenos' that is, why are the nerves to the diaphragm called phrenic?

DR. ALTSCHULE: The phrenic area is just above the kidneys where the diaphragm is located, a structure innervated by the phrenic nerve. The Greeks and others assumed that the phrenic area was the seat of thought or at least feelings. The Bible mentioned this also, and eventually the root phren as used as a prefix or a suffix began to refer to the head because presumably this was the seat of emotions and thought. The result was such terms as phrenology, which is the study of the bumps of the head.

To summarize, the schizophrenia concept did not spring fully developed from the furrowed brow of some genius. There was a long gestational period whose beginning we may arbitrarily date in 1683, with the writings of Thomas Willis (187). He used the designation fatuitas, which clung for a long time. After Hartley, in the mid-18th century, wrote his book (187) on the association of ideas, physicians applied this view of mental activity to fatuitas. Crichton in 1798 was apparently the first to do this in writing and Pinel soon followed. However Pinel called the disorder, démence, and in his book, published in 1809 (146) he made a formal statement concerning its symptoms which we still use today. Haslam gave us the earliest full clinical description and I have therefore suggested that those who reject the term schizophrenia might call the cluster of symptoms the Pinel-Haslam Syndrome.

DR. KETY: Did Haslam give it a name?

DR. ALTSCHULE: He neither gave it a name nor referred to the use by others of fatuitas in England and démence in France.

This history I have given you shows how ideas about the disorder became solidified by the mid-19th century.

Therapeutic Developments

We might note in passing that there was a ferocious dispute about the value of centrifuge therapy at that time, Connolly attacking bitterly the use of this treatment. He strongly advocated milieu therapy, all this almost a century and a half ago. The shock treatment used was centrifuging the patients.

DR. KETY: Dr. Benjamin Rush was one who advocated centrifuge therapy for a while because he thought it changed the cerebral circulation.

DR. ALTSCHULE: There were different concepts and techniques advanced by those who practiced the procedure. Some used the head-up position and others used the opposite. They clearly had opposite effects on the cerebral circulation.

DR. SNYDER: How does it work? Is it something like the swings in an amusement park?

DR. ALTSCHULE: No, it was a real human centrifuge. The patient is placed in it and he is swung around at an increasingly high velocity. It resembled the human centrifuges used today in experiments on aviators.

Our comments on the history of the schizophrenia concept should not fail to emphasize the views of some physicians that the syndrome whatever it was called then, was primarily a disorder of affect.

We must note also that Morel, who mentioned the term dementia praecox, evidently thought little of it. In fact his writings on the subject have given him praise and prominence he did not deserve. However, other authors took up the term, notably Pick, and when Kraepelin wanted a term to designate the multifaceted syndrome that commonly led to deterioration, he chose the term dementia praecox, to distinguish it from senile dementia. It should be emphasized that what is today called "return to an inadequate premorbid personality after a psychotic episode" was called deterioration by Kraepelin.

DR. BLISS: Would you make some comments about Kraepelin's pathological descriptions of the condition.

DR. ALTSCHULE: In those days about half the patients in the mental hospitals had syphilis and perhaps another large group had

Pathological Findings

different kinds of degenerative brain diseases or tumors. Some of the clinical descriptions that Kraepelin and others reported at the time were really cases of syphilis. Kraepelin's descriptions of somewhat lesser changes are quite difficult to make out because, of course, we all know that the brain tissue is very difficult to fix. Specifically the brains appeared normal in many of his patients excepting the elderly ones in which there was evidence of atrophy. However in 1890, pathology was still undeveloped and there were still no good methods for fixation of brain tissues. A surprising thing is that the German pathologists up to World War I rarely used the microscope.

DR. CANCRO: I would like to say one thing in defense of Bleuler's statement that although there may be some degree of recovery in chronic schizophrenia, patients in remission are never quite the same as they were before. In other words, there is no such thing as a full restitution.

DR. ALTSCHULE: That is true.

DR. BLISS: When reading Kraepelin's original papers, one is impressed by the fact that he had a large number of cases. He also quotes contemporaries whose series were also very large. So when he talks about dementia, he had a large amount of data to substantiate his views.

DR. ALTSCHULE: Yes, but what he calls dementia and deterioration leaves us with some doubt as to his definition.

DR. BLISS: We have to assume that he was working with a chronic type of hospitalized patient.

DR. ALTSCHULE: He also saw a considerable number of patients privately and in a typical clinical setting. However, Sullivan and others of his contemporaries described the symptoms of schizophrenia somewhat differently. I am not disparaging classifications because we could not deal with these things if we did not have a word that covered a cluster of signs and symptoms. But they are also subject to change. That brings me to another philosophical point and that is, what is classification all about? It is probably a matter of individual taste and I refer you to the mathematician-author, Lewis Carroll (28).

DR. SNYDER: Did Kraepelin put all these signs and symptoms of catatonia, hebephrenia together under his term dementia praecox--did he include catatonics?

DR. ALTSCHULE: Yes, but he identified the various syndromes specifically. However he had no etiology because ideas about the etiology of mental diseases have not changed much in over two thousand years of medical historical records. One finds the same common circumstances like overwork, underwork, oversexed, undersexed, too much money, too little money and so forth. Too much worry, too little love. Those comments are found with the earliest recorded remarks, but of course in the early days we still did not have four letter words. But I can quote you from a book that was written in 1610 on the subject of melancholia. "It is discouraging to note that the disease has not changed in recorded history." (18). The manifestations of mental disorders are the same as they were recorded in the Bible or before. In essence what they say is that everything has changed but the disease remains the same.

CHAPTER II - The Nature of Schizophrenia

Participants: Drs. Altschule, Bigelow, Bird, Bliss, Cancro, Gjessing, Kety, Lipton, Wolf, Wyatt.

DR. BLISS: The literature on schizophrenia is voluminous, contradictory, and confusing. It has been dominated by two antithetical views represented by the persons of Kraepelin and Freud. Kraepelin perceived schizophrenia, named by him Dementia Praecox as a disease of the brain, which appeared in many forms both blatant and subtle. It had a histopathology, ran in families, and was therefore presumably genetic (109).

In contrast Freud, although he gave passing respect to biology, devoted his attention to the psychology of schizophrenia. In the Schreber case (56), he brilliantly organized his subject's delusional thought, and showed that bizarre, irrational beliefs, not only made intrapsychic sense but were also a response to intolerable conflicts. From this has emerged the concept that schizophrenia is a learned behavior, predicated upon early maladaptive experiences, a behavior that appears when intolerable problems leave no alternative for coping except psychotic tactics.

As a derivative of these two antipodal positions, the one biological and the other experiential, is the diplomatic compromise that schizophrenia represents the confluence of a biological defect and pathological experience. According to this psychosomatic orientation, the schizophrenic has a biological defect, which is expressed if the individual has been prepared by early deleterious experiences.

The Schizophrenias and Latent Schizophrenia

That the brain should have its diseases should come as no surprise. Every organ in the body is heir to a wide spectrum of disorders. Some are infectious, others traumatic and some may be genetic. The variety is great and their causes are multiple. Furthermore, since the brain has unique functions, including higher mental activities, it would not be unexpected to find specific diseases which distort mental and emotional behaviors. It seems likely that schizophrenia is one of these disorders.

Next, one would anticipate that schizophrenia would have its blatant, obvious and severe forms, as well as its more subtle, less severe forms. This is certainly true for nephritis, cirrhosis, arthritis and myriad other diseases. In fact, one would guess that the less severe forms would far outnumber the graver disorders. There is an enormous literature on these flamboyant disorders, which seem to be related to schizophrenia. They have been subsumed under a variety of names including pseudoneurotic schizophrenia, spectral disease, schizoidia, ambulatory, borderline, etc., schizophrenia. Unfortunately there is no marker, as yet available, to identify these cases with certainty. A summary of this literature would involve a monograph in itself, since there is much evidence that these more subtle schizophrenias do exist in large numbers. A recent review of monozygotic twins discordant for schizophrenia indicated that 43.5% had neurotic or characterological abnormalities, females being more prone to the former and males to the latter (31). Heston (84) has commented that the clinical delineation of this borderline schizoidia group is most unsatisfactory since these individuals do not fit neatly into our present diagnostic groupings. He refers to Slater's (169) description of their traits and indicates that 45 to 50% of the parents, siblings and children of schizophrenics have been found to be overt schizophrenics or to exhibit what can be considered latent schizophrenic disorders.

Table I lists descriptive words and phrases used by relatives of schizophrenics in describing their abnormal but non-schizophrenic relatives (169). The inference which emerges reinforced by much circumstantial and clinical evidence, is that the genetic trait, when it is not flagrant, may display itself in a variety of psychopathologies, although the exact forms are not yet defined. One can only guess that if the prevalence of severe schizophrenia is 1%, the prevalence of borderline cases might be as high as 5 or 10%.

A final comment about the "disease" schizophrenia seems appropriate. The history of medicine indicates that when a disorder is discovered, it rarely remains a single disease. This is exemplified by hemolytic diseases, cancer, infections, diabetes, nephritis, jaundice and the like. Even in psychiatry we are beginning to discover that there are many depressions -- unipolar, bipolar and schizoaffective disorders which appear to be distinctive. It would indeed be wonderously strange if schizophrenia proved to be one disease. It seems much more likely that it will have many etiologies, and that multiple genetic defects will eventually be discovered, amidst a variety of causes.

DR. WOLF: In considering the term schizophrenia in its historical perspective the early students of the process were interested in describing it, in trying to find out what happened; then came the emphasis on possible causes; only relatively recently has the focus turned toward mechanism. The older terms "weakness of associations, inappropriate behavior" in more modern terms may be explained as evidence of defective coupling. A number of investigators demonstrated that when ACTH is given to schizophrenics, their eosinophils fail to drop or be depressed. Others have demon-

Dissociation as a biological concept. Disorganization of adaptive responses

TABLE I

PARANOID TRAITS: Suspicious, sensitive, sullen, touchy, grouchy, morose, resentful, unforgiving, difficult, quarrelsome, self-conscious, jealous, litiginous, critical, takes things the wrong way, has rows with all the family, does not get along with people, makes heartless accusations.

ECCENTRICITIES: Giggly, opinionated, pedantic, narrow-minded, meticulous, obstinate, humorless, rigid, conventional, conceited, superstitious, prudish, cranky, miserly, foxy, precise, brusque, verbose, circumstantial talker, little-minded, full of facts, learned but incompetent, old-fashioned, routine-bound, has bizarre ideas strongly held, spiritualist, believes in self-cure by hypnotism.

LACK OF FEELING: Impassive, cool, calculating, placid, hard and stingy, disciplinarian, unsympathetic, cold, slack, unscrupulous, withdrawn, very sane, little feeling, unkind and selfish, unconcerned about a debt.

RESERVE: Shy, serious, staid, haughty, snobbish, studious, unforthcoming, independent, taciturn, unsociable, quiet turns, seeks solitude, quiet old stick, exceedingly reserved, no give and take, never reveals his thoughts, absorbed in scientific pursuits, one friend only.

ANERGIC: Feckless, dependent, tired, slack, unreliable, subservient, a poor thing, unable to work and health gave way, separated from family and tramped, no initiative or money sense, neglected family and went downhill.

strated that in schizophrenics, the cold pressor test is unreliable. For example, when the hand is immersed in cold water an increase in blood pressure is expected to occur. However, schizophrenics may put their hand in cold water but not develop this pressor effect.

DR. ALTSCHULE: Well, I would challenge that.

DR. WOLF: Water intoxication associated with inappropriate antidiuretic hormone secretion has been described in association with agitated depressive psychosis (149). Water intoxication is very difficult, if not impossible, to induce in a normal person no matter how much water he drinks. I have observed water intoxication in three schizophrenic subjects who drank water excessively. They became unconscious with very low serum osmolality. However, when these same subjects were relatively improved, unconsciousness associated with very low osmolality could not be induced. This again suggests some sort of a defective coupling, possibly failure to turn off the antidiuretic hormone. Adolph Meyer pointed out some years ago that one can identify a schizophrenic through conversation because you do not feel that you can identify with or convey a sense of understanding or sympathy for him because you feel as though you are not plugged in or loosely connected with him. All this falls under the general rubric of "defective coupling", a growing concept of brain biochemistry which we will be discussing as we go along and which may provide a foundation for observations.

DR. ALTSCHULE: Perhaps so but I think there are many concepts to consider. We can show that many of the stress phenomena found in schizophrenia are quite normal.

The sort of disturbed, exaggerated or deficient adaptations that Dr. Wolf described are not specific for schizophrenia but these and other disturbances may be found in otherwise healthy individuals under great stress or after injury.

My first example is from a study of 17-ketosteroid excretion, made a quarter of a century ago. The patient was admitted with an initial diagnosis of compulsive-obsessive neurosis, but it was soon changed to schizophrenia. The patient was chosen because she could be relied upon to give us accurate urine collections. Her excretion of 17-ketosteroids was high, but fell as she improved with ambulatory insulin therapy. However, a rise in 17-ketosteroid output such as this untreated patient exhibited was in no way specific, and was merely an evidence of stress. What was the nature of the stress in this untreated schizophrenic patient? We do not know.

Another abnormality found in psychotic patients is a decreased utilization of the keto-acids derived from glucose, including pyruvic, oxalo-acetic, and alpha ketoglutaric acids. The levels of these substances rise abnormally after glucose is taken. It is important in doing glucose tolerance tests of this sort that the patients be prepared with a high carbohydrate intake for three days beforehand and that they be at absolute bedrest during the entire test. The same abnormalities occur in patients with a wide variety of recognizable stresses - fever, blood loss, trauma, myocardial infarcts - as well as some metabolic disorders, e.g., thiamine deficiency and hyperthyroidism. What is the nature of the stress the schizophrenic patients are suffering? We do not know. Nor do we know the mechanism of the abnormality. In psychotic persons the abnormality is not reversed by thiamine or by thioctic acid. It is possible that it is due to a deficiency of the largely neglected chromium nicotinates. When patients go into remission the metabolic abnormality disappears. It recurs a few days before clinical relapse.

DR. KETY: How much of that picture could be produced in normals by the intramuscular adminstration of ACTH?

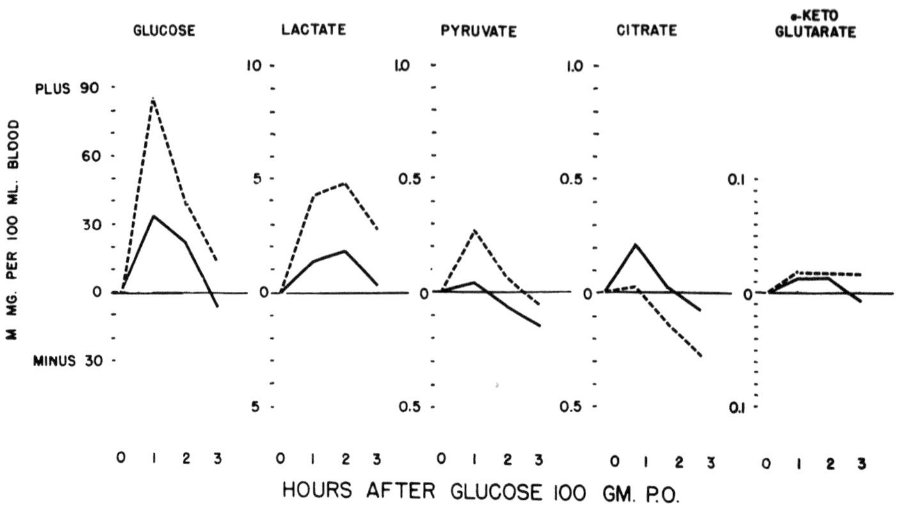

FIG. 1. Effect of ACTH on carbohydrate metabolism (average of 3 subjects). Solid lines indicate values obtained after 100 gm of glucose by mouth in control period. Dotted lines indicate values similarly obtained after three days of ACTH, 100 mg per day.

DR. ALTSCHULE: That is exactly what we did. In Fig. 1 you can see individuals who are physiologically normals given ACTH and others are what we later term as schizophrenics. We got exactly the same picture in these normals as in the psychotic patients given nothing. Recently Richard Veech, at Saint Elizabeth's Hospital in Washington, got much the same changes in liver metabolism in rats that are stressed.

Another phenomenon seen in psychotic patients is the excessive rise in blood glucose level when they are given both ACTH and glucose. However this hypersensitivity to ACTH is not specific; it is found in persons with spontaneous or induced Cushing's Syndrome.

Similarly the low blood eosinophil count and the insulin resistance found early in the course of schizophrenia, are also nonspecific and merely indicate the existence of stress. The same is true of the elevated serum copper level. It simply reflects the presence in a schizophrenic of stress that may have gone unrecognized. Fig. 2 shows that certain oxidation products of epinephrine and other catecholamines have a hemolytic effect. We thought we discovered it but a search of the literature shows that it was known many years before. The red blood cells of patients with schizophrenia show an excessive amount of hemolysis when incubated with the oxidation products of epinephrine. However this is likewise not

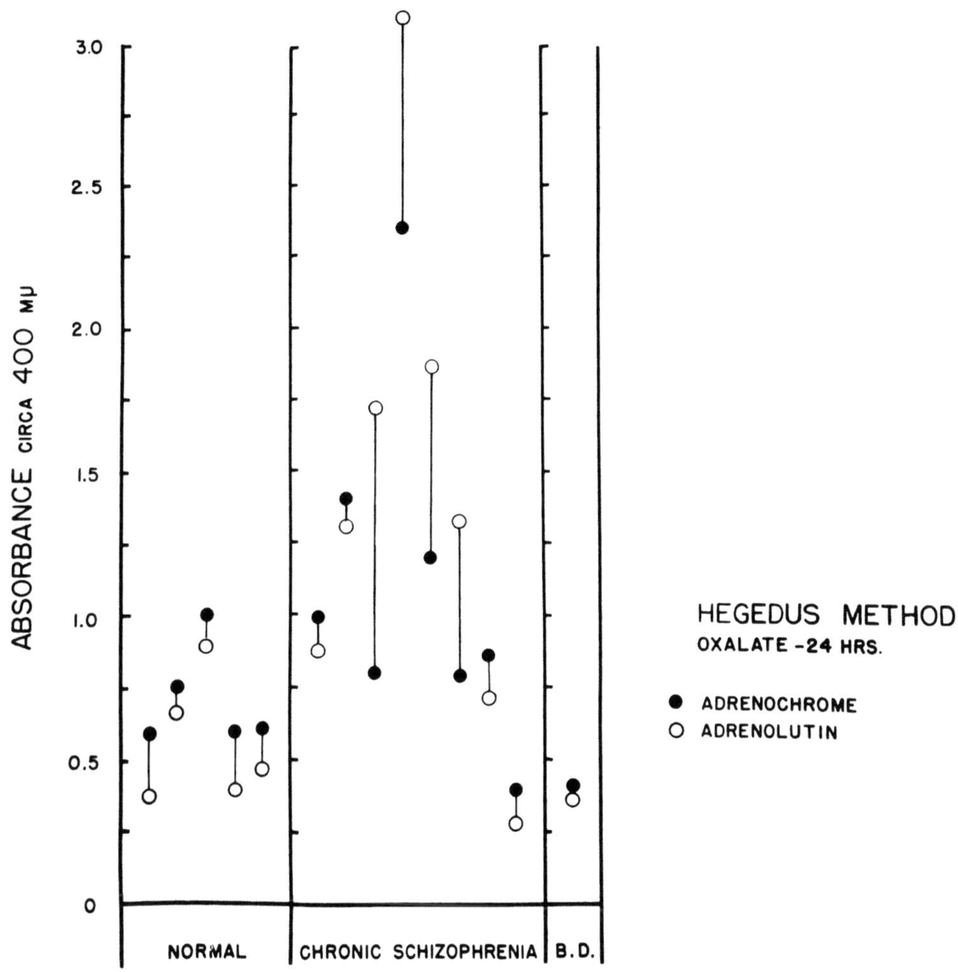

FIG. 2. Hemolytic effect (absorptive circa 400 mμ) after incubating oxalated blood with adrenochrome or adrenolutin (B.D. signifies chronic brain damage).

specific, as it was also found in wounded men at the Chelsea Naval Hospital and at the Boston City Hospital. We can conclude from this that there is something in the plasma, or missing from the plasma, of stressed people which is characteristic of stress and also found in patients with schizophrenia.

Still another test we studied was the effect of the injection of phlorhizin. This prevents the reabsorption of filtered glucose in the renal tubules and hence causes a great loss of glucose. In normal persons this loss is compensated, and the blood glucose level remains normal, for six hours at least. On the other hand, in schizophrenics there is a considerable fall in the blood glucose level. These patients mobilize glucose slowly from their body stores. However studies in other non-psychiatric illnesses give the same results. Here we have still another test that is not specific for schizophrenia.

DR. KETY: Was liver glycogen depleted?

DR. ALTSCHULE: No, one reason is that we force fed them glucose for three days prior. Some other people doing the same experiment found in biopsies of the liver an abnormal amount of glycogen. Thus, stressed individuals may not mobilize glycogen into glucose.

Furthermore, it has been reported that most schizophrenic patients have a zero plasma level of ascorbic acid upon admission to the hospital. Such individuals require about ten times as much ascorbic acid to get a measurable plasma level of the vitamin. However, that is also true of a number of other sick people.

None of these tests should be regarded as diagnostic with any degree of specificity for schizophrenia.

DR. KETY: Did you say that the hemolysis is precipitated by the oxidation products of catecholamine?

DR. ALTSCHULE: Yes, and this is true in some patients with cancer, with myocardial infarction, or with trauma as well as in schizophrenia.

DR. WOLF: Should it not be added to this that the non-specificity idea is related to the fact that the homeostatic mechanisms or the physiologic responses in sick schizophrenics are disturbed and disordered in a fashion similar to that encountered in severe wounds or other stresses.

DR. ALTSCHULE: I agree.

DR. WOLF: The mental and behavioral features of schizophrenia may be induced under circumstances of severe military stress, a syndrome that is distinct from chronic schizophrenia. That is to say that healthy individuals under great stress may become confused, disorganized, may hallucinate, develop paranoid ideas and may even become catatonic. One sees the physiological as well as the behavioral and mental manifestations that are conventionally associated with schizophrenia. These all disappear when the stress is removed. The point about non-specificity should not be limited to the physiological manifestations, therefore, but should include all of the manifestations we identify as schizophrenic.

Schizophreniform Manifestations occurring under stress

DR. KETY: Yes. The various phenomena that are seen in schizophrenia may be seen acutely in patients under stress, but the whole picture of classical schizophrenia is rarely if ever seen merely as a result of stress.

DR. WOLF: During World War II in New Guinea, we saw temporary psychoses that were altogether indistinguishable from catatonic schizophrenia.

DR. ALTSCHULE: The clinical picture of schizophrenia may also be simulated by drugs. At the time of the Crusades there was a wave of drug abuse throughout Europe. Also during the period of the Napoleonic conquest of the Middle East, the soldiers brought back with them hashish, henbane, hyoscyamus and atropine and, of course, the opiates. Then another wave of drug abuse spread over Europe.

DR. WOLF: Miss Tracy Hughes has raised an important question that we should try to settle before moving into other facets of this subject, "What does it mean to a patient when the label 'schizophrenia' has been affixed?" That is to say, what homogeneity is there in this syndrome and what common ground do those of us here at this table have in visualizing this. We have repeatedly emphasized the absence of specificity of the various manifestations of schizophrenia. What are the criteria by which we recognize it? I refer not only to the disorganized mental manifestations of the disease, but also to the physiological behavior of the individual. Where is the solid ground?

The significance of the label, Schizophrenia -- Diagnostic considerations

DR. CANCRO: Until such time as you have an independent biologic measure which is constant and replicable, any effort at a scientific classification is going to be deficient. The clinical strategy is one in which you observe and are on the lookout for, specified symptoms which will permit the data to be fitted into a particular diagnostic pattern. Today if you took a random group of psychiatrists and presented them with a group of 100 psychotic individuals with the request that they identify those with the diagnosis of schizophrenia, there would be considerable disagreement and low reliability. On the other hand, if the psychiatrists were first instructed to use a specific set of criteria in establishing the diagnosis of schizophrenia, and if they, in fact, applied those criteria, the final result would show good agreement as to which individuals are schizophrenic. There is good agreement amongst any group of psychiatrists regarding the presence or absence of specific symptoms or signs; therefore, if you want to use such symptoms and signs as criteria you will wind up with a patient group in which there is considerable reliability and agreement as to diagnosis (193). However, if the psychiatrists are instructed to use their own personal criteria in their own individual fashion, we obviously shall wind up with a wide variety of opinions as to what constitutes schizophrenia and which amongst these patients are schizophrenic (166).

DR. KETY: I would like to comment because everything that Bob Cancro has just said is true and I agree with it heartily. However, I think that Dr. Altschule in developing his concept really described schizophrenia for what it is: a syndrome which, of course, is simply a form of cluster analysis which physicians have been using for centuries before the advent of computers. It is nothing more than that but still it is very useful. These cluster analyses form syndromes useful for communication and research and from these come further discoveries of further differentiation that lead us to more and more specific etiology. As a matter of fact, the word insanity was, I suppose, applied to the original syndrome and then in the course of the study and practice of psychiatry it was recognized that there was a form of insanity which led to progressive neurological deficit of a certain kind and ultimately general paresis emerged. Eventually, from the study of general paresis it was learned that people with that cluster of symptoms also had had a history of syphilis. And from this, the origin of the possible relationship of syphilis to the mental process was divined and finally came the discovery of the spirochete as the causative agent of syphilis. But I think we ought to respect the right of the person who first describes the syndrome and describes it in the manner in which he first saw it. We should not arbitrarily add other concepts to it unless it can be shown by an independent group that these criteria, that these other subgroups really fit into that syndrome as it was originally described. Schizophrenia is not a disease. It is simply a phenomenological collection of symptoms which seem to hang together and in the Kraepelinian sense, most psychiatrists who have been well trained will recognize it just about as well as any other physician who has been trained to recognize most other syndromes. And not a single one of these features can be considered totally characteristic of schizophrenia. Indeed no single symptom is really pathognomonic of any syndrome.

DR. CANCRO: While I agree with Seymour Kety that no single sign is pathognomonic of a syndrome, it can be a useful strategy to restrict the diagnosis of the syndrome to certain signs and to exclude others. It has been shown that if you restrict the diagnosis, for example, to those individuals who show a typical thought disorder, it tends to make the population far more homogeneous in real and useful ways (22). Therefore these efforts can be helpful in defining the syndrome more carefully.

DR. GJESSING: The large group of mentally retarded individuals resemble schizophrenia, in many respects. Among the mentally retarded it has been possible little by little to demonstrate the association of metabolic errors with a given set of symptoms. At the moment we know of about 40 different types of inborn errors of metabolism associated with brain damage. In the case of galactosemia, fructose intolerance syndrome and tyrosinemia, we have three separate diseases with the common feature of failure to thrive. Their symptoms are essentially similar. They have in common aminoaciduria, pathological changes in the liver and kidneys, and may die during a slight infection. In each case, correction of the specific biochemical defect is not only life saving but essentially restores health. This suggests that there may be many biochemical determinants in schizophrenia.

DR. LIPTON: It will help us to look at mental retardation and the progress that has been made in the understanding of that condition because originally it was treated as a syndrome. In mental retardation a specific illness can be identified in about half of the severe cases but we are still left with about 50% at the syndrome level. If we consider moderate or mild degrees of mental retardation, the figures become much worse. There are still about 80 to 90% of cases at the syndrome level. I am not very optimistic in thinking about schizophrenia, that we shall soon be able to generate specific illnesses out of both the severe and mild schizophrenias.

DR. KETY: The example of mental retardation is a particularly good one because if one plots the frequency of human intelligence, one does not get a simple gaussian distribution, but a bi-modal distribution. The major distribution may represent the scatter to be expected in a polygenic trait like intelligence, but the second mode at the low end of the scale represents a collection of different diseases which have in common a low I.Q.

DR. LIPTON: If we consider schizophrenia as a relapsing disease featuring more than one or two episodes, then I would agree completely that this is a bi-modal phenomenon. Those individuals who display only one transient psychotic episode, then would not be considered schizophrenic in the Kraepelinean sense. They would fall into the category of acute schizophrenia, a stress disorder that is apparently reversible. Nevertheless some psychiatrists would still call the condition 'Schizophrenia'.

DR. KETY: But that is the point that I was making. It is not right to preempt Kraepelin's nor Bleuler's diagnostic terms which they clearly applied to a chronic disorder.

DR. LIPTON: Those who have redefined schizophrenia to include not only acute psychotic episodes but also certain types of chronic deviant social behavior would consider schizophrenia as fitting on a Gaussian curve of behavior.

DR. CANCRO: Inherent in this approach is the danger that it can lead to what I call the "everyman theory", which states that we all can become schizophrenic if submitted to the right stresses. The proponents of this approach can always explain our failure to become ill with the argument that you have not had the right stresses yet and maybe tomorrow will be more productive. That stance is not very constructive. In other words, there has to be some unique construct in the conception of the syndrome that would make it identifiable. It is only by its uniqueness that we are able to identify it. Certainly if you include chronicity as a parameter, you will find much greater homogeneity in your identified individuals. For example, nowadays you may find a group of adolescents at the age of 16 or 17 who have been "freaked out" as a result of taking virtually the entire U.S. Pharmacopeia but then after a few days when all these drugs have been excreted they do not look anything like the traditional schizophrenics. They are nice adolescents with warm dispositions and effective personalities. They should not be considered schizophrenic.

Schizophrenic behavior as a normal variant

Psychotic behavior may be encountered in the absence of drug taking. What some call schizophrenia may be a feature of acceptable behavior in certain cultures.

DR. KETY: I might add that I think the American school of psychiatry is, to a large extent, responsible because it has so broadened the concept of schizophrenia that it has included in it people who are showing deviances which are part of the normal distribution. Some do not even realize that there is a severe debilitating syndrome -- which really is schizophrenia.

DR. WOLF: Is such deviant behavior the basis for Szasz' The Myth of Mental Illness? He stated that schizophrenia is not an illness but that its manifestations are part of a general distribution.

DR. KETY: Szasz' arguments are based largely upon the undesirable social consequences of the use of the term schizophrenia. Laing uses another kind of non-sequitur which is that no one has demonstrated a pathological or biochemical lesion in schizophrenics, therefore none exists. That is a complete non-sequitur.

DR. LIPTON: But Szasz' position, you might say, differs depending on whether it is his public position or his private position because he is very politically sensitive. I suspect he would deny it but some years ago when I asked him, "Tom, do you really believe there is no mental illness?", he replied in words to this effect, "Of course I don't believe it. Sure, there is mental illness but that is not the point. Consider the large number of people who are not mentally ill, who are being treated as if they were. That is what I am arguing against. These are the big majority of people."

DR. BIGELOW: I agree that politics must play a part in Laing's position. There is no doubt that ill equipped, ill staffed and inhumane hospital conditions have an effect on the schizophrenic's behavior as well as on his relative comfort and, to some extent, on the type and degree of pathology which he will express. And so to that extent, one must reckon with Laing's and Szasz' position. Their position can have a productive consequence. It will not negate the existence of the illness but their observation will remind us that these are still human beings and that we have the responsibility to see that they be treated as such.

DR. WYATT: Dr. Lipton made the point that the greatest progress in the study of mental retardation, came through the study of those patients furthest from normal. Using that analogy, should we be studying schizophrenia in the sickest group of people?

DR. LIPTON: Trying to work with truly chronic patients who have been schizophrenic for fifteen years or more might not be very rewarding. They may be in a stable state where it is difficult to find out anything.

DR. BIGELOW: Neither should we select acutely ill young people having their first -- and possibly last -- episode for inclusion in any small study of new therapeutic approaches. The expectable recovery rate is high and good treatments are available for most. It is not reasonable to withhold these treatments. In like manner an extremely "burned out" group of chronically hospitalized subjects is scarcely attractive. Although we have no sure prognostic criteria or clinical marker points which establish the potential reversibility or irreversibility of any of the components of the schizophrenic syndrome, many seem after a time to be suffering more from the consequences of a previous illness than from an active ongoing process. The finding that many long stay patients may be taken off phenothiazine tranquilizers without noticeable effect supports strongly the view that an active disease process is no longer present.

On the other hand, all experienced clinicians have observed instances of apparent recovery in long stay patients. These dramatic remissions seem to occur consequent to significant interpersonal or physical events, and are often short lived. Still, the fact that ECT, Amytal, high fever, or major illness can be associated with temporary remission does argue against there being irreversible structural alteration as part of the schizophrenic process. Perhaps that which is irreversible is the psychic damage secondary to the chronic failure implicit in the schizophrenic's life.

The best group for study, I propose, is the chronically acute one. By this I mean patients whose apparent illness remains active as expressed by the presence of delusions, hallucinations or other perceptual dysfunction, and some evidence of anxiety.

Doing research with this type of patient has its hazards. Although in our research design we attempt to set up only one variable at a time, others creep in whether we like it or not. There is no such thing as behavior independent of environmental input.

DR. LIPTON: I agree fully with Dr. Bigelow. Ed Stainbrook once declared that schizophrenics learn to schizophren. I suspect that for the truly chronic patient this may be the case, just as prisoners learn to live in prison and they cannot adapt to living outside prison, so they go back in. From that perspective, I think that trying to work with chronic patients who have been hospitalized schizophrenics for 15 years or more might be very difficult especially if the work is in the area of therapeutics or mechanisms. They may be in such a stable state of adaptation that it is difficult to find out anything. On the other hand, if you select for study the acute schizophrenic, you may find that the patient does not have true schizophrenia. It might be better to look for patients who have had two or three relapses out of the hospital but are still "relatively fresh cases." Those, I think, would be the ideal cases to work with if one is trying to do either metabolic or therapeutic studies. Burned out cases, whether they are functionally burned out or organically burned out, may be too far gone for such studies. Equally one is in trouble if you are really not sure that it is schizophrenia. The acute psychotic episode, which is not schizophrenia, is where lots of the mischief has come, particularly in therapeutic claims. The easiest way to get positive therapeutic results is to choose a case that does not have the illness in the first place. I think this holds for the megavitamin results in many cases. Mainly they treat what they call "phase one" cases. Such patients may very well not have schizophrenia to begin with.

DR. BLISS: There may be many schizophrenic diseases, some being chronic, others acute, and some subtle. I would expect a spectrum. I do not believe that the argument about acute and chronic schizophrenias has been resolved.

Distinction between Acute Schizophrenic Reaction and Schizophrenia with acute onset

DR. LIPTON: We have all seen cases, some of them have been reported in the literature, where observing one acute episode and being obliged to make a diagnosis, you would have to call it schizophrenia. But a week later the symptoms are gone and there is no recurrence over a twenty year period of observation. In such cases we presume that the etiology is a devastating personal or military experience. One almost has to call this type of case a schizophreniform psychosis. It looks like schizophrenia at the moment of examination but it has a benign outcome.

DR. KETY: It does not really look like schizophrenia if we remember that two of the cardinal features of schizophrenia are chronicity and an insidious onset which such acute cases do not show. Rather one should describe them as psychoses resembling schizophrenia in cross section. It is certainly true that many patients who ultimately develop chronic schizophrenia begin with an acute psychotic episode. In Denmark Welner and Stromgren followed a population of schizophreniform psychoses, which in the U.S. we would call acute schizophrenia, and found that a third of such patients ultimately developed other characteristics of schizophrenia including chronicity (184). In the individual with an acute onset there may be a subgroup who are showing the initial manifestations of schizophrenia. But they should not be labelled schizophrenics until chronicity is evident.

DR. CANCRO: Unless we can, on some clinical or other grounds, differentiate those acute cases who will go on to chronicity and typical outcomes from those acute cases who do not, we are at a disadvantage. At this time we can do a lot better retrospectively than prospectively. If they were all initially designated as schizophrenic we would be wrong 35% of the time.

We cannot say if this is the first acute episode in what will be a chronic deteriorating course or if it is a reaction to some intercurrent life event which is not likely to recur again. It is very difficult to differentiate these cases from each other since we do not have the markers at this time to help us out.

DR. LIPTON: Dr. Kety, you are correct in correcting me. The initial definition dealt with a chronic illness. And there is no way to know that it is chronic without looking at it over a period of time. You are also correct in that one-third of the acute episodes go on to repeats. I think we need another term besides chronic, an intermediate term, to describe repeated episodes which do not develop into the "back ward" state hospital type case. But Dr. Cancro raises an important question. I have sometimes wondered, given one of the St. Elizabeth's patients who possibly had chronic, "back ward" type schizophrenia, for twenty years, if we discovered the equivalent of a penicillin for that illness, would we ever recognize it in those patients? My guess is I do not think we would. I really think that there are institutionalization syndromes which are learned in the process of living in a back ward. Those cases might be so irreversible that we could not do anything with them.

Reversibility of the Schizophrenic Syndrome

DR. KETY: We have heard the argument often but is it really true? What happened when one could treat paresis or pellagra? What happens to schizophrenics with chlorpromazine? I think it is remarkable how readily these patients can and do adjust to being well.

DR. LIPTON: They can get better, the paretics, those with the real tabes dorsalis did get better, but they were hardly ever well. They had residual organic damage. The patient with ten years of chronic pellagra also had neurologic damage. Nicotinic acid certainly made him better but he had residual damage. I suggest that ten to twenty years of hospitalization may leave the patient with behavioral and social deficits that might make biological improvement difficult to detect.

DR. CANCRO: We do know that we are able to discharge many back ward chronic schizophrenics who were institutionalized for many years prior to chlorpromazine. Many of them did show recurrence, social breakdown syndromes, inappropriateness, things of this sort. Their adjustment left something to be desired, but the adjustment was enhanced by the chlorpromazine. Obviously one cannot disentangle how much of the symptomatology was institutionalism and how much was, in a sense, a residue that was not sensitive to chlorpromazine.

There is some evidence that chronic patients who are handled in noninstitutional settings do not show quite the same syndrome as patients who are handled in institutional settings where there is very little socializing, very little human contact, etc.

DR. LIPTON: If you had a choice of any number of patients to work with and you were working on pharmacotherapy, what kinds of patients would you choose to work with in order to have the greatest chance of success?

DR. CANCRO: Those acute cases who are young, preferably drug using, with good premorbid personalities.

DR. LIPTON: Those would include the pseudo cases.

DR. CANCRO: That is right. Well that is in answer to your question. You asked which ones would I pick to maximize therapeutic success.

DR. LIPTON: No, I mean for reliable, meaningful pharmacotherapeutic results. Not to become a hero or to get rich by curing schizophrenia but rather to understand the illness. I think I would feel that if you take the most severe cases, that is where you have the greater chance of picking up the metabolic defect.

DR. BIGELOW: Following this argument, take multiple sclerosis. What do you suppose would happen there? One would be studying end stage. In this case there is obvious pathologic destruction of tissue.

DR. WYATT: You might have a chance of identifying the virus if there is such an animal in a chronic multiple sclerotic, rather than in a person with a first time retrobulbar neuritis.

The Search for Specific Features toward the Identification of Schizophrenia as a Disease

DR. KETY: I do not really think there is a single answer. There are a number of strategies that one can use. Obviously one would like to find the most cardinal and most classical examples of the syndrome to work with but if one chooses only people who are most severely affected, one does run the risk that in them the metabolic defect if there is one has burned itself out and what one is seeing are the scars only. One might use another kind of strategy, selecting patients with a heavy genetic load in their families. They would at least be a homogeneous group that one could dissociate from people who do not have a very strong genetic load. One might also study those who are very intimately related genetically to schizophrenics such as the discordant monozygotic twins.

DR. GJESSING: It might be possible, as a strategy in the study of "schizophrenia" to select cases that appear as clusters in a family. Starting at that point one might attempt to identify common biochemical or other potential genetic markers.

Alternatively one might detect a shared responsiveness in the group to a certain drug. Such clues could open doors to the more rational study and classification of schizophrenia. Refsum's disease or heredopathia atactica polyneuritiformis, is an example of the identification of a specific disease entity from a large group of patients with closely similar clinical manifestations. This disorder was first recognized in 1938 and later described by Professor Sigvald Refsum in Oslo in 1945 and 1946. The main clinical features of this disease are: (1) an atypical retinitis pigmentosa with hemeralopia and concentric constriction of the visual fields; (2) a chronic polyneuritis with progressive paresis and muscular atrophy in the distal parts of the extremities; (3) Ataxia and other cerebellar signs; (4) Increased protein content of the cerebrospinal fluid without increase in the number of cells. The clinical course is characterized by marked exacerbations and remissions, or stationary phases over several years with more or less progression.

In 1963, when gas chromatography was available, Klenk and Kahlke in Germany discovered large amounts of peculiar fatty acid, namely phytanic acid, in serum and in different organs. The abnormality was found only in patients with Refsum's disease.

Later it was shown that the accumulation of phytanic acid in patients with Refsum's disease was due to dietary phytol, a constituent of chlorophyll and phytanic acid, found in significant concentrations in butterfat and in the tissues of ruminant animals. The accumulation of phytanic acid from the diet was due to an enzyme deficiency of α-hydroxylation of phytanic acid.

By giving a diet low in phytol and phytanic acid, the level of serum phytanic acid was lowered to normal values. Simultaneously both subjective and objective signs of clinical improvement were noted. No relapse has been observed in patients on this diet observed over long periods.

All this seems clear cut. Up to 1973 it had been possible to collect about 40 patients from the general category of heredo-familial ataxias, by using phytanic acid as the pathognomonic sign. But the story is still not finished.

Patients with similar syndromes but with normal concentrations of phytanic acid have been found. This is consistent with the possibility that yet not described defects can lead to chemical abnormalities, perhaps by similar pathophysiological mechanisms.

This illustrates that it took about 35 years, with a specific technique to select out the distinct entity, Refsum's disease, from the general category of heredo-familial ataxias.

How much more complicated must it be within the field of the spectrum of the schizophrenic syndromes with mainly behavioral and psychological symptoms.

DR. WOLF: There is a purported relationship of schizophrenia to artistic creativity. What can you say about that?

Creativity among Schizophrenics

DR. ALTSCHULE: Whereas a good many creative writers have had manic depressive psychoses, the number of recognized, widely praised authors who have had schizophrenia is very small.

DR. CANCRO: Let us look at some data. The Heston study (83) on the adopted away children showed there was more musical talent among the offspring of schizophrenic women than amongst the offspring of normal controls. There was also a Russian study (79) which showed that the offspring of schizophrenic women when given a test of creativity -- they gave them objects and they were supposed to come up with unusual usages -- were able to come up with more unusual usages for these common objects. That was obviously a limited test of creativity. Another interesting study came out of Iceland within the last few years, in which they did two things (102). They took a group of schizophrenics and looked at their relatives to see how many made it into "Who's Who in Iceland". Then they took the individuals from "Who's Who in Iceland" and looked at their relatives to see what the prevalence rate looked like and they did find a clear statistical association between this indicator of creativity and schizophrenia. And while what Mark Altschule says is true about the writers, there are a number of writers who come to mind who have schizophrenic children. And there are a number of creative scientists who come to mind who notably enough are Nobel Laureates, who have had schizophrenic offspring.

DR. ALTSCHULE: You cannot judge creativity in certain categories like writing. You can judge it in painting. You can judge it in music. You may perhaps be able to judge it in science although I doubt that.

DR. CANCRO: But there are studies and data are available. It is not wholly a matter of "I knew a patient once and he was very imaginative." It is not totally anecdotal. There is data to suggest a linkage between creativity in terms of the scientific achievement and literary achievement.

DR. ALTSCHULE: That is exactly what Haslam said in 1809 but he put it much more eloquently (76).

DR. KETY: Well, one can see similarities between schizophrenic thinking and certain types of creativity. I have the notion that the creative individual is one who can develop a large number of off-beat, novel, non-commonplace associations but the creative scientist, and the creative artist, has another ability which is a very critical ability to reject most of them and to hold on to one or two that make sense in other ways. It may be that they share with the schizophrenic this ability to make novel associations, but scientists must have the critical ability to reject them. I know a number of creative scientists who are on lithium, but only one very good scientist who was institutionalized as a schizophrenic.

DR. CANCRO: I think Dr. Kety has raised a very good point when he said a part of the creative process is the ability to see new relationships. On the other hand, he also tempered that by saying that it is necessary to be able to reject postulated relationships which are not productive and to sustain involvement with a new insight long enough to test its value.

DR. BLISS: If one assumes that some schizophrenias are genetically determined, then it would not be amiss to conclude that intelligence, artistic ability and the latent schizophrenic trait could co-exist in some talented individuals. Under these circumstances the latent thought disorder might contribute to creativity, since imaginative innovations might ensue. For example, I am told James Joyce had a schizophrenic daughter.

Epidemiology

DR. BIRD: Is there any evidence on racial or geographic variation in schizophrenia? I had the impression that there was an increased incidence among American Indians.

DR. CANCRO: There are some old studies that appeared years ago in some of the Southern medical journals saying that schizophrenic blacks were different.

EDITORIAL COMMENT

While the epidemiological studies leave a great deal to be desired, they have not adduced any clear evidence of racial differences. The study of Simon et al. indicates no differences between whites and blacks when strict diagnostic criteria are applied (167).

DR. KETY: The only verified incidence I know of, and Dr. Gjessing would know this better than I, is a population in Northern Sweden studied by Böök (13). This population, in a very remote northern province of Sweden, had a 3% incidence of schizophrenia, three times higher than the rest of Sweden.

DR. GJESSING: That is right. This population was of special genetic interest because of its social homogeneity and the inbreeding. The clinical syndrome was predominately catatonic. Sheperd and co-workers, in collaboration with the World Health Organization, carried out introductory studies all over the world with uniform criteria for "Schizophrenia." They found the same incidence in these different groups all over the world. There seem to be no significant differences due to geographical or social conditions (165).

Geographic and temporal variability in incidence and prevalence

DR. WOLF: What about time to time variability in prevalence of schizophrenia in Scandinavia, where epidemiological information is very good? So many diseases have been reported to be very fluctuating, more now than then. Has there been a sustained level of prevalence of schizophrenia or has this varied with the historical changes and social changes that have come about in Scandinavia?

DR. GJESSING: It has been very stable. According to Professor Ornulf Odegard in Oslo, "Various statistical investigation have shown that the incidence of mental disorders does not increase during wars. This is particularly true of psychoses, but (with the exception of certain acute reactions in the actual zone of combat) even of the neuroses as well." (138).

"During the German occupation of Norway 1940-1945, the number of first admissions for psychoses to psychiatric hospitals decreased about 15% in the diagnostic categories 'schizophrenia and reactive psychoses'. Readmissions were unaffected, and so were first admissions from senile, epileptic and other organic psychoses. A minor part of the decrease is a consequence of the loss to our statistics of all war-time psychoses in Norwegian seamen, soldiers and political refugees abroad."

Odegard has also investigated Hospitalized Psychoses in Norway; Time trends 1926-1965 (139). He states: "A case register comprising all admissions to psychiatric hospitals in Norway since 1916 has been examined for possible time trends during this fifty year period. Only minor changes were found to have taken place in the total incidence of the psychoses and in their marital, occupational and geographic patterns of distribution."

On the other hand, the epidemiology is very well investigated in Iceland by Professor Tomas Helgason (80). He studied all Icelanders born during 1895-1897, who were alive in Iceland on December 1, 1910. The life history of each individual was obtained for as long a period as possible or until July 1, 1957. Due to the very extensive sources of information sufficient data could be obtained on 99.4% of the 5,395 probands. The expectancy for schizophrenia in per cent was 0.57 - 0.69 for men and 0.9 - 1.02 for women, the higher rates derived from estimates based on cases with certain as well as cases with uncertain diagnoses. There was no significant difference between expectancy of mental disorders in Iceland and in Denmark (where K. H. Fremming did a similar study on Bornholm) when the expectancy is estimated in the same way for comparable age groups and for comparable diagnostic groups in both countries (55).

DR. WOLF: This is of great interest because we are used to other diseases varying enormously from time to time. This was so beautifully said by René Dubos in his "MIRAGE OF HEALTH" (42) and those of us who were house officers in the thirties and early forties remember the almost endemic level of rheumatic fever in the northeastern part of the United States and then the prevalence of rheumatic fever decreased well before the now abandoned prophylatic chemotherapy was practiced. Instead the incidence of rheumatic fever increased greatly in an area where there had previously been very little, namely Mexico City. We are accustomed to such variability in the case of many diseases, cancer of the stomach for example and many others. So it is very startling and striking, that the incidence and prevalence of schizophrenia appears to be monotonously uniform from geographical area to geographical area, as you point out, and from time to time through turbulent differences in social structure.

DR. CANCRO: This fact has been used as an argument to play down the role of social, cultural, psychological, and experiential factors, and has been used as an argument for genetic transmission. I think what is perhaps even more interesting and instructive concerning the conclusion that can be drawn from these observations, is the low reproductive rate seen in this disorder for many years.

The increasing reproductivity of schizophrenics

Schizophrenics had been reported to produce few offspring up until the advent of chlorpromazine and the out-of-hospital treatment era which is only since 1955. Many people thought that there was an endocrine abnormality in schizophrenia since they did not have children. They drew an obvious inference first, but forgot that if patients are locked up in a hospital, and if they do not have intercourse, they have a very low reproductive rate. But this has changed. The reproductive rates have recently climbed nearly to the level of the general population (46). The question will be, "Is there a change in the prevalence rates in the future?" Will there be a change post-1975, or post-1980, when there will be twenty years or more of chlorpromazine experience and where the offspring of schizophrenics start to come into the risk period? Does this Scandinavian village represent a kind of test of that question? If there is an isolated small population with considerable inbreeding, will this lead to a doubling or tripling of the prevalence rates? Is this something that we can expect as we move towards the increased reproduction of the schizophrenic population? These questions were raised by Seymour Kety several years ago. I remember as a speaker you discussed it during the meeting we had in the Spring of 1969.

Problems in evaluating Epidemiologic Data

DR. KETY: I think we have to be very cautious in drawing any conclusions at all from prevalence rates of schizophrenia. One is not dealing with an element which has an objective, definable criterion, and the truth of the matter is that the prevalence rates or the lifetime incidence which have been recorded in Europe and Japan are still showing a wide spread. In Germany during the thirties the incidence was reported as .5%. There was one study in Switzerland where it was 1.25% and there are studies in between, and we oversimplify that by saying, "Well, the incidence is about 1%."

But the incidence of schizophrenia varies tremendously depending upon the criteria one uses, and upon what the hospitalization rate is in the country. In Denmark, the incidence of schizophrenia is much higher in Copenhagen than it is in the Provinces, not because the true incidence is any different, but people out in the farms who are schizophrenic get along and those who are in Copenhagen and schizophrenic do not get along and so they come into the hospital. So that I really do not think that we can draw any conclusion that schizophrenia has a uniform incidence throughout all time or all places. We just do not know. In general there is a variance, it fluctuates around 1%, but you could change that figure of 1% from .3% to 1.5% simply depending on how you define schizophrenia, and in America we could easily get a figure of 2% for what is called schizophrenia, whereas in a European country the figure would be low.

DR. CANCRO: I think the best example of this is the Soviet Union where they get rates of 4 and 5%. They are not terribly troubled by these rates because, by their criteria, it is constant. I think it still is an interesting fact that if times are good or bad, depression, boom, bust, war, or peace, within a particular culture, you do not see, as long as the diagnostic criteria are fairly constant, a fluctuation with social events.

DR. KETY: You could have a 100% increase in the incidence of schizophrenia and never know about it, because it is a relatively rare disease.

DR. CANCRO: You are working with small numbers.

DR. KETY: That is right.

DR. WOLF: I think, however, that these studies Dr. Gjessing referred to did use the same criteria for diagnosis. The workers concluded that there was a rather uniform geographical distribution (165).

DR. GJESSING: They examined populations in five or ten countries around the world.

DR. KETY: Oh, yes, but the thing that was striking about that study was that qualitatively schizophrenia looked the same in these different countries, but it would be very difficult to infer an incidence figure from their data because how does one compare an incidence in India or in Africa with an incidence in Norway or in Washington, D.C., when the techniques of ascertainment are so different.

DR. GJESSING: Prior to the study they agreed upon the diagnostic details, and trained the people in these different countries to use the same diagnostic methods.

DR. KETY: But they did not go out into the community and round up 100 people or 1,000 people and interview them.

DR. GJESSING: Well I do not know the details. Shepherd gave a talk about it and I got the impression that it was an astonishingly small variation from country to country.

DR. KETY: But you are simply saying that if you do not see the patient you do not make the diagnosis. That is right. But even in Denmark the incidence in Copenhagen is different from the incidence in the country.

DR GJESSING: Yes, but you know there are some people in the country who are not so schizophrenic that they have to be in the hospital. They can live and work on the farm during a certain period of the year.

DR. KETY: Right.

DR. GJESSING: But otherwise there is a register that collects all cases from all over the country, and they know even if he is at home, he is under the care of a doctor. He has been diagnosed and registered.

DR. KETY: No, no. It is the registered cases that are lower in the country. In our interview study we found that a certain number of people that we diagnosed Kraepelinean schizophrenia were not known to the psychiatric register because they had never seen a psychiatrist.

DR. GJESSING: Not known to any hospital, not to any doctor at all?

DR. KETY: Exactly, right!

DR. GJESSING: That can happen.

DR. KETY: Now if that can happen in Denmark, it must be happening in India and in Africa to an even greater extent. So that one cannot really figure out the true incidence of the illness and it could still be fluctuating very markedly and we not realize it.

CHAPTER III - Neural Substrates

Participants: Drs. Altschule, Bigelow, Bliss, Cancro, Kety, Wolf.

DR. BLISS: The genetic nature of some schizophrenias seems to be well established, although genes do not explain the entire problem. Furthermore, there is increasing evidence that "learning" to be schizophrenic does not seem to be the answer.

There is no doubt that schizophrenics live in a social setting; they do not exist in a vacuum. It is a common clinical experience to learn from some schizophrenics that their acute breaks were preceded by conflicts and impasse. Bowers (15) quotes such statements from patients as, "I had nowhere to turn," or "There was no way out." In many cases, life threatening conflicts are perceived but neither flight nor fight seem possible.

However, these are the stresses which presumably overwhelm the predisposed. The critical consideration is the nature of other factors, besides the genetic one, which may play a vital role in some cases. The thesis which I will present is that brain damage may, in fact, be another variable in some cases, which has generally been slighted, at least by American psychiatrists.

Neurologic features of Schizophrenic syndromes

I have long assumed that the schizophrenic must experience extraordinary sensations to conclude that he is hearing voices, seeing visions, and being influenced by external forces or when he indulges in echolalia or echopraxia. The following observations speak to the possibility of brain damage or some metabolic disturbance, akin to a "psychedelic" experience.

A number of investigators have begun to re-examine the schizophrenic experience anew (15, 29, 53, 120). Table II lists some of the experiences which are reported by early schizophrenics. It should be noted that all schizophrenics do not relate these. They appear at least early in the disorder, seem to be episodic, and many schizophrenics have a wide variety of them. They are subjective, occur in that black box called the mind, and thereby lack the scientific respectability that some demand. Unfortunately, so much that is schizophrenia resides in this domain.

The symptoms described in Table II bear a striking similarity to those experienced by individuals suffering from lesions in the frontal, temporal and parietal lobes of the brain. As such they are akin to the aphasias, complex sensory disturbances and disturbances

TABLE II

I have so many thoughts at the same time, that I can't put them together.
My mind blocks -- goes empty.
I can't control my thoughts. They go by themselves.
One thought leads to so many thoughts -- that it is confusing.
I am much more aware of things that are happening around me.
Things look blurry, fuzzy, unclear.
Things suddenly become larger or smaller or taller and taller.
Things and people look split up, like a picture that is torn into bits, and then put together again.
Things that I see look strange and different.
When I look at objects they look flat -- without any depth.
Colors look much brighter than normal.
My arms and legs do not seem to be a part of me. They just move on their own.
Sometimes I feel confused -- feel I don't know where I am or who I am.
When I speak the words don't come out right.
I get so many thoughts, all together, that I can't keep them straight.
When I talk now I put words in the wrong order.
When I listen to people, their words come into my head all right, but they seem to get scrambled in my brain.
Sounds seem louder and clearer.
Sounds seem softer and less clear.
Everything tastes the same.
Peculiar odors come to me at times.
I feel like someone else.
When I move - it feels awkward - my movements are not automatic.
Moving is like a motion picture. If you move the picture in front of you changes. The rate of change in the picture depends on the speed of walking. If you run you receive the signals at a faster rate.
I am virtually blind at these times and can't move properly because there are so many things coming into my eyes that I don't know what's what.
Often I have to go through two or three things in my head before I find the thought I want--words I don't want come out--not the correct word--not the words I wanted for the meaning I wanted to give.

of the special senses which neurologists study. Furthermore, some are experiences which seem to reflect aberrations in focusing and attention perhaps attributable to disturbance in the reticular activating system. Thus there is a need for more precise study of the schizophrenic's remarkable mental experiences, uncontaminated by psychodynamic bias.

Brain Damage - Evidence from Pneumoencephalograms

One of the dogmas of schizophrenia is embodied in its designation as a functional psychosis. It has been assumed that schizophrenics have no demonstrable neuropathological changes, an assertion that remains open to question in view of most arresting reports scattered throughout the literature of dilated ventricles in many schizophrenics -- particularly the refractory, severe cases. These reports suggest a high incidence of generalized or focal atrophy in this population. The findings in eight of these studies are specified in Table III.

All studies were done as clinical investigations using conventional clinical criteria for normal ventricles and were predicated upon pneumoencephalograms. One study done by Haug (78) published in 1962 as a monograph, impressive for its attention to details and adherence to strict diagnostic criteria, will be summarized in detail.

He studied patients admitted to the Dikemark Hospital in Oslo. After a careful review of pneumoencephalographic literature, he selected precise standards for normal, borderline and the abnormal. In fact, his were conservative standards, since his upper limits of normal size for the ventricles was a little higher than those ordinarily used by neuroradiologists. The purpose was to avoid the error of including normal variants as abnormal (Table IV).

A standard procedure was followed in all cases. Thirty ml. of air was injected in all tests and the focus film distance was 70 mm. Psychiatric diagnoses were made independent of pneumoencephalographic findings, and all measurements were done by the same neuroradiologist without knowledge of diagnosis. In a number of cases, pneumoencephalograms were repeated. If the patient's status had remained unchanged, test-retest reliability was excellent. Cases of known severe alcoholic abuse and severe head trauma were excluded from the group. The schizophrenics were hospitalized chronic patients with a mean age of 41. The author was cognizant of and took account of the expected enlargement in ventricular size which occurs in the middle years. Table V summarizes the findings in schizophrenia.

By all criteria, including widened sulci, 84 of the 137 cases (61%) were judged abnormal.

TABLE III

AUTHORS	POPULATION	FINDINGS
Jacobi & Winkler (93)	19 young, severe schizophrenics	18 (95%) abnormal
Lemke (114)	100 chronic, severe schizophrenics	50 (50%) dilated ventricles
Moore et al. (126, 127)	71 schizophrenics sick 1/2 - 18 yrs.	25 (35%) dilated ventricles
Huber (88)	190 schizophrenics of every variety and stage	131 (69%) dilated ventricles
Borenstein et al. (14)	134 schizophrenics	118 (88%) abnormal
Bratfos & Sagedal (16)	40 schizophrenics	17 (43%) abnormal 8 (20%) uncertain (atrophy) 15 (38%) normal
Young & Crampton (199)	36 schizophrenics (referred for neurological consultation)	24 (66%) atrophy 6 (17%) uncertain 6 (17%) normal

TABLE IV

VENTRICULAR STANDARDS

Radiographic Dimensions in mm.

Region	Normal	Probable Dilatation	Definite Dilatation
Lateral Ventricles (septal caudate line)	= or < 14	15 - 16	= or > 17
Third Ventricle	= or < 9	10	= or > 11
Temporal Horns	= or < 4	5 - 6	= or > 7

TABLE V

VENTRICULAR SIZE IN SCHIZOPHRENICS

No. of Cases	Schizophrenic Diagnosis	Normal Ventricles	Probable Dilatation	Definite Dilatation
45	Hebephrenic	13	10	22
37	Catatonic	10	7	20
52	Paranoid	18	10	24
3	Unclassified	1	2	0
137 Totals		42 (31%)	29 (21%)	66 (48%)

EEG's done on many cases revealed a significant increase in abnormalities among schizophrenics with dilated ventricles. On the other hand, Haug found a significant number of patients with dilated ventricles who had normal EEGs.

The type of dilatation was examined in a subseries of 43 cases of schizophrenia. The findings are listed in Table VI.

The conclusion drawn from these and other data was that schizophrenics tend to have diffuse, bilateral dilatation, although there are a significant number of exceptions.

Table VII indicates the relationship between clinical deterioration and cerebral atrophy. There was clearly a positive relationship between the degree of clinical deterioration and the presence of a definite dilatation of the ventricles. Of the 101 schizophrenics with onset of the disease in adolescence, 28 had a disastrous course with early deterioration. Of this group, 71% had dilated ventricles and only 7% had completely normal pneumoencephalograms.

Haug also analyzed his data to determine whether treatment with ECT or insulin might have contributed to his findings. Fortunately, he had a large number of patients who had not received these therapies. Their pneumoencephalograms were contrasted to those of the shock treated patients (Table VIII). There was absolutely no difference between the two groups, suggesting that ECT and insulin were not contributing to the ventricular dilatation.

Clearly these findings of a high incidence of cerebral atrophy in the schizophrenics would come as a surprisè, if not a shock, to most psychiatrists. These studies are virtually unknown to American psychiatrists, and such a proposition would be considered preposterous by many. It may be, in fact, that these studies are in error but they cannot be ignored.

Neuropathology and Neurological Signs

I have not reviewed the neuropathological literature in detail but I do know that it is large and contradictory. It should be resurrected and critically examined. There is a recent report by Colon (32) on "schizophrenic dementia" in which losses in cerebral mass and cells are described. A problem in all of these studies is the one of incorrect diagnoses, but one must ask what the correct criteria for the diagnosis of schizophrenia are.

So-called soft neurological signs have been noted in many studies of adult schizophrenics (104, 145, 152, 180), adolescent schizophrenics (81, 82, 112), and in childhood schizophrenics (50, 67) and have been reviewed by Pincus and Tucker (145) and Cam-

TABLE VI

VENTRICULAR SIZE IN VARIOUS SCHIZOPHRENICS

Schizophrenic Diagnosis	Diffuse Dilatation Bilateral	Diffuse Dilatation Unilateral	Focal Dilatation
Hebephrenic	12	2	3
Paranoid	6	1	3
Catatonic	10	2	4
TOTALS	28 (65%)	5 (12%)	10 (23%)

TABLE VII

CLINICAL DETERIORATION VS. VENTRICULAR SIZE

Clinical Status	Normal Ventricles	Probable Dilatation	Definite Dilatation
No deterioration	19 (68%)	5 (18%)	4 (14%)
Probable deterioration	10 (37%)	9 (33%)	8 (30%)
Definite deterioration	9 (20%)	6 (13%)	31 (67%)

TABLE VIII

RELATIONSHIP OF SHOCK THERAPIES TO VENTRICULAR SIZE

Population	Normal Ventricles	Probably Dilated	Definitely Dilated
Treated with ECT and Insulin	17 (36%)	10 (21%)	20 (43%)
Untreated	20 (37%	11 (20%)	23 (43%)

pion and Tucker (20). Only a few will be cited as representative.

Subtle Neurological Signs

Minor signs include astereognosis, agraphesthesia, extinction during bilateral simultaneous stimulation, bilateral hyper-reflexia, defects in coordination, disturbances of balance and gait, mild movement disorders, abnormal motor activity, chorei-form movements, and anisocoria. Goldfarb (67) has found that schizophrenic children have significantly more soft signs than do controls. Fish and Alpert (50) noted neurological changes in infants of schizophrenic mothers, related to patterns of arousal, decreased motor responses and poor muscle tone.

Recently Tucker et al. (180) studied 58 consecutive admissions to an acute inpatient psychiatric service. The Goldstein-Sheerer object sorting test and the sensorimotor portions of the Halstead-Reitan battery (finger agnosia, finger tip writing, tactile form recognition and tactile form performance test) were administered. Sixty per cent of those diagnosed as schizophrenic showed impairment, in contrast to 21% of the non-schizophrenics.

These findings in many studies, although "soft" by neurological standards, begin to be persuasive when they are related to the many other findings of central nervous system dysfunction culled from different segments of the literature.

Schizophrenia in the Known Brain Damaged

The literature relevant to the relationship of schizophrenia to antecedent known brain damage is muddled. Much of this is summarized by Achte et al. (1). They also report their study of 3,552 men followed over a period of 22 to 26 years who incurred war brain injuries. 317 or 8.9% became psychotic during this period. Unfortunately their subcategories of psychosis are difficult to reconcile with our nomenclature. I have simply listed in Table IX those which might be considered by some as "schizophrenic." This is indubitably the largest series of brain damaged individuals who have been studied in this way over a long period. All were soldiers who served in the Finnish Army during the wars of 1939-1945. Whether selection factors excluded some of the mentally ill from the military is unknown to me, but presumably some of the males at risk for schizophrenia must have been eliminated from the Army. In any event, the incidence of "schizophrenic like" psychoses is high compared to the anticipated rate of ±1% in the general population.

Birth Injury

In a series of studies Pasamanick and Knobloch (142) demonstrated that prematurity and complications of pregnancy are associated with brain injury. These may lead to fetal and neonatal death, cerebral palsy, epilepsy, mental deficiency, behavioral

TABLE IX

"SCHIZOPHRENIAS" IN A BRAIN DAMAGED POPULATION

	Number	% of 3,552
Schizophrenic psychosis	76	2.1%
Paranoid psychosis	70	2.0%
Epileptic psychosis	56	1.6%
Affective psychosis	37	1.0%
Reactive psychogenic psychosis	7	0.2%
Others	10	0.3%
TOTALS	256	7.2%

disorders, reading disabilities, infantile autism and tics. The most obvious insults appear to be due to prolonged anoxia in the course of complications of pregnancy such as toxemias and maternal bleeding. Furthermore, these complications of pregnancy and the resulting neuropsychiatric disorder are aggregated in the lower strata of society.

Other factors which may contribute to these reproductive casualties have been found. They include maternal-fetal blood type incompatibilities, nutritional deficiencies, endocrine disorders, radiation, maternal age -- very young and older mothers -- drugs, a variety of infections, cigarette smoking relating to prematurity, and emotional turmoil during pregnancy.

The question is whether some schizophrenias are also in part reproductive casualties. The evidence is clearly not conclusive but many observations suggest this possibility. In Table X some of this information is summarized. To be noted is the high incidence of perinatal mortality and mental deficiency in offspring of schizophrenics as well as the high rate of mental deficiency in some schizophrenic populations. Furthermore, a number of studies indicate a greater cluster of complications of pregnancy in the chronic mentally ill and the mothers of schizophrenic offspring. Moving from the perinatal period one is impressed by the mortality rate in childhood reported for children of schizophrenics. Furthermore, it should be remembered that schizophrenia is concentrated in the lower socioeconomic class, where these reproductive casualties are the highest. What does emerge from the confluence of circumstantial

TABLE X

"REPRODUCTIVE" CASUALTIES

	Retrospective Data About Adult Schizophrenics
Stabeneau et al. (175)	A study of 11 monozygotic twins discordant for schizophrenia. The schizophrenic member had a lower birth weight; and a higher incidence of cyanosis, infantile colic, feeding problems, fractures and severe illnesses.
Stabeneau and Pollin (174)	A study of 2 monozygotic twins discordant for schizophrenia with same birth weight. Index twin had CO poisoning and almost died. Other index twin had Rocky Mountain Spotted Fever at age 3 1/2 and almost died.
Robins, L. (151)	Schizophrenics had more severe infectious diseases in the first 2 years of life, more hearing problems, more motor problems, more tics and mannerisms, more eating and sleeping disturbances.
Nameche and Richs (137)	Normal births in less than 40% of cases. They had slower motor development, poorer coordination, delayed development of speech and speech problems.
Woerner et al. (192)	Schizophrenics had more complications of pregnancy and delivery than their siblings.
	Mental Deficiency
Kallman (100)	6.8% of the children of schizophrenics were mentally defective (General Rate = 1.5%)
Hallgren and Sjogren (74)	10.5% of schizophrenics were mentally defective.

TABLE X continued

	Infant and Child Mortality with a Schizophrenic Parent
Reisby (150)	In a population of 322 offspring of 132 schizophrenic mothers, 9.3% of the children died in childhood.
Sobel (172)	Perinatal mortality = 8.1% Malformations = 3.2%
Kallman (100)	Perinatal mortality = 5.4 - 7.9% (General Population = 3.6%) If one parent was schizophrenic 44.5% of their children died before the age of 5.
Canavan and Clark (21)	If one parent was schizophrenic, 18% of their children died in childhood.

High Risk Studies Children of Schizophrenics Studied Early in Life	
Fish and Alpert (50)	13 infants born to schizophrenic mothers were placed in foster homes immediately after birth. At age 2, four had "soft" neurological signs.
Mednick et al. (122) Mednick (123)	Neonates had lower birth weight. 70% suffered pregnancy or birth complications (anoxia, prematurity, prolonged labor, placental difficulties, etc.) Infants had retarded motor development (holding up head, walking with support).
Sameroff and Zax (159)	An increased number of perinatal complications which was a function of the chronicity and severity of the mother's mental illness - rather than schizophrenia.

TABLE X continued

High Risk Studies Children of Schizophrenics Studied Early in Life (cont'd)	
Rosenthal and Nagler (156)	Neurological test items that differentiate children of schizophrenic parents from controls are facial asymmetry, fine motor coordination, visual perception, left-right orientation and auditory visual integration (true for children under the age of 11 years).
Erlenmeyer-Kimling (45)	High risk children, particularly at ages 7, 8, and 9 show greater impairment than controls on right-left identification, rapid alternating movements, reading, ability to wink and choreiform movements.
Mednick and Schulsinger (125)	Lower birth weight, more retarded motor development (age at which hold up head, walk with support, sit, stand, etc.).
McNeil and Kaij (121)	No increase in perinatal complications.

evidence is the greater likelihood that some schizophrenics, due to the mental illness of parents, their habitat, socioeconomic circumstances, complications of pregnancy, neglect, nutritional deficiency and other factors, incur brain damage.

Since most individuals with brain damage do not become schizophrenic, it may well be that one needs the genetic predisposition as well. Although a few observations may not prove a theory, occasionally apt ones may strongly support it. Stabeneau and Pollin (174) reported two pairs of monozygotic twins discordant for schizophrenia. In both instances the member of the twinship who became schizophrenic had experienced an illness, which would be expected to cause brain damage -- in one case, carbon monoxide poisoning and in the other, a severe encephalitis. One might infer that there was a latent tendency to schizophrenia in both pairs of identical twins, but that only the two with the brain damage developed the manifest form.

If this hypothesis were true, then certain predictions would follow. First, one would predict that the genetic intensity of the schizophrenic trait would be milder in the discordant pair, than it is in the concordant pair. From this, one would expect fewer overt schizophrenias in the first degree relatives of the discordant pairs.

In the only paper which directs itself at this question, Rosenthal (154) indicates that there was almost a total absence of schizophrenia in families of discordant twins, whereas the illness occurred in about 60% of the families of concordant twins. Next, one would predict that offspring of identical twins discordant for schizophrenia would breed an equal number of children with psychopathology, since the twins are genetically identical. This finding has been reported in one study (49).

Returning to Table X, there is also evidence of "soft" neurological signs early in some "high risk" children or in some children destined to become schizophrenic. This can be explained as evidence of early brain damage due to reproductive insults, or as some unexplained manifestations of the genetic trait.

Electroencephalographic findings

When one considers the heterogenity of schizophrenics, the number of extraneous variables including drugs, exercise, anxiety, diet, hospitalization, physical inactivity, etc., which distinguish them, as well as those schizophrenics who may have undiagnosed brain damage, the plight of the biological investigator is evident. It is not surprising that electroencephalography has not been definitive. There is a long list of studies by many techniques including spectral analysis, many of which have revealed variation from the normal (39, 63, 68, 89, 92, 115, 118, 171). The latest by Itil et al. (92) is cited simply because the technology was excellent. They did not attend to the many other variables which may or may not have been critical. In their computerized period analyses of the EEG of chronic schizophrenics and normals they first demonstrated that the patterns of individuals remained very stable from week to week and month to month. They found that chronic schizophrenics had a greater number of slow waves, a greater amount of fast activity, fewer alpha waves, a greater average frequency and frequency deviation, and a lower variability in amplitude than did the normal subjects.

In another study the same group compared the EEGs of 31 children of schizophrenic mothers to those of 50 controls (91). High risk children had variations from the normal, similar to those found in the adult schizophrenics.

It was concluded (60) that there is a sizeable (20-60%) subgroup of drug free schizophrenic patients whose EEG records are characterized by low voltage, fast, choppy activity and a low index of alpha rhythm. This is a pattern seen at the cortex of excited animals and has been interpreted as evidence of hyperarousal or anxiety.

Whereas the EEG has many obvious limitations it was hoped that the technology of averaged evoked potentials would bring refinement and definition to these problems. Unfortunately, these problems have multiplied. Shagass (163) has recently reviewed the subject. He comments that the results obtained by comparing various psychiatric groups with respect to averaged responses to unpaired stimuli have been either negative or, when positive, have been difficult to confirm. In part, the problem has been one of methods since there must be careful control of pupillary size and the visual stimulus. He also concludes that all investigators who have compared evoked-response variability in schizophrenic patients and nonpatients, have found greater variability in schizophrenics. There are other details but it seems evident that this electrophysiological technique has, as yet, failed to illuminate the dilemma of schizophrenia.

From these various studies, I would suggest the following conclusions:

1. The contribution of genetics to schizophrenia seems well nigh indisputable.

2. There are probably many schizophrenias.

3. For every typical case of schizophrenia, there are probably 5 or 10 latent cases.

4. There is much evidence to support the opinion that some schizophrenics have brain damage. In some cases, this may represent degenerative disease of the CNS which has been misdiagnosed. On the other hand, brain damage may add an incremental disadvantage to individuals with schizophrenic genes. It seems likely that brain damage will permit latent schizophrenia to become manifest; and may convert mild schizophrenia into a severe form.

5. Offspring of schizophrenics, for a variety of reasons, appear more liable to incur brain damage.

6. There is even the possibility that schizophrenias are neuropsychiatric diseases with a neuropathology that remains to be discovered.

DR. BIGELOW: Some years ago Randy Rosenthal and I did a study using autopsy material, just measuring with a planograph the various structures that could be measured. We measured about ten or fifteen different areas, such as hypothalamus, cortical thickness and the thickness of the corpus callosum. To our amazement, rank ordering it, there was no overlap between the thickness of the corpus callosum in the controls versus the schizophrenics. It was the only significant finding. The corpus callosum was larger in schizophrenics (155).

Changes in the Corpus Callosum

DR. WOLF: What about the histology?

DR. BIGELOW: We are now planning a more detailed study to include histology.

DR. WOLF: That is in the works, is it?

DR. BIGELOW: Yes. Laitinen reported that one could reduce the anxiety of schizophrenics by the production of lesions in the genu of the corpus callosum (111). I have seen no subsequent reports.

DR. WOLF: Let me turn the discussion to the localized disturbances in the brain in relation to schizophrenia. We heard about the corpus callosum being enlarged. One would speculate, I should think, that the disturbances in the brain such as infiltrative tumors or encephalitis, localized possibly predominately in the corpus callosum, might simulate the schizophrenic process. What do we know about that? During World War II some individuals with hepatitis were admitted to the psychiatric ward in pre-icteric phase, because of manifestations typical of catatonic schizophrenia. Then after a few days, as icterus developed, the psychotic manifestations subsided. As far as I know, from our own experience, these individuals recovered completely. But as far as the psychopathology is concerned, they had the manifestations of schizophrenia. This is also a common clinical experience in encephalitis. Sometimes it is very difficult to make the differential diagnosis early in the game. Personally I have seen some instances of infiltrating gliomas with schizophrenic-like behavior.

Neurologic lesions simulating schizophrenia

DR. KETY: There are many neurologic disorders which have sometimes been confused with schizophrenia. We have often thought that the temporal lobe must have something to do with schizophrenia because so many temporal lobe lesions are confused with schizophrenia. Some hold that retrospectively they can dissociate neurologic ailments from schizophrenia. For example, Teuber (178) showed that lesions in one or another part of the brain very often produced one or another schizophrenia-like symptom. But when he was asked, is there any neurologic disorder which is indistinguishable from schizophrenia, he said "No."

More interesting kinds of observation I think relate to those instances in which the true illness is confused with schizophrenia by a good psychiatrist. In Wilson's Disease, for example, Scheinberg, an expert in this malady, has noted patients may be diagnosed as schizophrenic before the diagnosis of Wilson's Disease becomes apparent. Interestingly enough, I learned recently that even in Huntington's Chorea, which is such a clear cut neurologic entity, the first case of this disease in a family is very often diagnosed as schizophrenia. It is only when other cases appear in the family the disease is no longer confused with schizophrenia. And then, of course, there is the case reported by Harvey Mudd, which is the first and only schizophrenic in which a clear cut enzymatic deficiency has been established. In this case the psychiatrist made a firm diagnosis of schizophrenia (130). But the patient had homocystinuria which was found to be due to a genetic deficiency of the enzyme methylene tetrahydrofolate reductase. Folic acid could correct the metabolic disturbance and cause the schizophrenia to disappear.

DR. CANCRO: Before you raise that question, let me add a word on one patient whom I saw several years ago, admitted to Medicine with ascites. The staff were having management problems because of his schizophrenic symptoms. The resident asked me to see him. The patient had an orientation impairment and the clinical picture just did not hang together. Although he had been diagnosed as schizophrenic at the local state hospital, he was transferred to the Medical Service when he developed ascites. By the time I got over to see him again the next day, additional history had been obtained and the diagnosis of Wilson's Disease was becoming established. So I think it depends to some degree on what criteria are used and how clear, for example, consciousness must be. Will you accept slight impairment of consciousness, or if there is any impairment must schizophrenia be excluded and some other diagnosis sought? If one uses relatively general criteria for psychosis, it is well to remember Ernest Gruenberg's classic statement that there are a finite number of ways of being "crazy." Independent of the etiology there are only a small number of ways that "craziness" can be manifested. Returning to this particular case, the diagnosis of schizophrenia was made. The patient was treated in the state hospital with phenothiazine which was just enough of an insult to his liver to precipitate the ascites and make possible the diagnosis of Wilson's Disease.

DR. WOLF: What effect did phenothiazine have on his behavior?

DR. CANCRO: It calmed down his agitation. A lobotomy would probably have produced the same result because agitation is a relatively nonspecific symptom. For a person who is agitated, even if it is an acute brain syndrome, phenothiazine will reduce the agitation.

DR. WOLF: But the other manifestations of thought disorder, disorganization?

DR. CANCRO: He still had the mild disorientation, the memory impairment, and the acute clinical picture was still there.

DR. BLISS: It is true that patients with medical or neurological disorders are often diagnosed as being schizophrenic. Many are acting crazy, and most medical people are not adept at making the subtle distinctions. On careful scrutiny most can be determined to be organic confusional states. However, a few present major diagnostic problems. More recently these have included some of those who have abused amphetamine or the hallucinogens.

DR. ALTSCHULE: I once saw a patient who was taking steroids for something or other. He developed typical catatonia including the psychological changes. Nobody believed the change was real. When the steroids were stopped the abnormalities cleared, only to reappear again with re-institution of the steroid therapy. Because of continued disbelief the steroid induced catatonia was replicated seven times.

We all see patients who are called schizophrenic by trained psychiatrists, who ultimately turn out to be suffering from Cushing's Syndrome, myxedema, pernicious anemia, thyrotoxicosis or severe sodium or potassium depletion after a diuretic. I cannot give you a percentage, but any one who sees many patients will run across some where the diagnosis of schizophrenia must be revised in the light of test results that indicate a metabolic or other physical aberration.

DR. WOLF: And some of these people might have schizophrenia too.

DR. ALTSCHULE: Yes, there is nothing specific about it.

DR. BIGELOW: Among a large number of patients referred as schizophrenics, we received one "guaranteed" schizophrenic, a 26 year old girl from New York who had been treated as schizophrenic in multiple state hospitals, clinics and private hospitals. When she got on the ward, it became evident that she had some organic basis for her ailments, because of a mild right congenital cerebral hemiatrophy. Her functional deficit was in the visual-spatial organization. Over the years since she was removed from school as prepsychotic at age 10 and sent to a residential treatment center, she accumulated quite a repertoire of schizophrenic vocabulary and behavior. In retrospect only, one can say that she had a lot of warmth underneath which did not fit in the schizophrenic picture.

DR. WOLF: She fooled everybody for quite a while.

DR. BIGELOW: For quite a while, yes. The point is that now some people are more alert for and aware of the anatomical reasons for developmental abnormalities in children, the various kinds of minimal brain damage are being sorted into more specific dysfunctions. Dr. Martha Denckla (40) found on examining our patient that her behavior matched very closely that of a large sample of children with behavior disturbances who have visual-spatial disorganization of a similar sort. Whether or not there is a link between that particular functional perceptual deficit and the disorders of brain circuitry in schizophrenia is an intriguing question.

CHAPTER IV - The Genetics of Schizophrenia

Participants: Drs. Altschule, Bigelow, Bliss, Cancro, Cohen, Kety Lipton, Snyder, Werthessen, Wolf, Wyatt.

DR. BLISS: Information about the genetics of schizophrenia is now persuasive, if not conclusive. Since this will be examined in detail by Dr. Kety, I will only briefly refer to the salient principles.

It now seems certain that there is an increased risk for schizophrenia in relatives of schizophrenics. Furthermore, the closer the genetic relatedness, the greater is the likelihood of schizophrenia (Table XI). This is comparable to the conclusions drawn by Sir Francis Galton in the 19th century, where he studied the likelihood of genius in relatives of a genius (57).

These data have been verified repeatedly, and there seems to be little doubt about their validity. They are further supported by studies of twins, concordant for schizophrenia. In essence it now appears that the concordance rate for dizygotic twins is about 10% whereas it is 40 to 50% for monozygotic pairs.

Since the variables of nature and nurture are not separated by these strategies, adoption studies have been employed to make the distinction (185). What emerges is the prepotency of genetic factors. Schizophrenic mothers produce offspring with a much higher than the normal expectancy rate for schizophrenia, even when the child is separated very early from the mother and raised by foster parents. The incidence of schizophrenia appears to be about 15% -- quite comparable to the figures for children raised by a schizophrenic parent.

TABLE XI

Expectancy of Schizophrenia
for Relatives of Schizophrenics
(Slater & Couie) (170)

Expectancy in the general population is ± 1%

	Certain Cases	Probable Cases
Parents	4.4%	5.5%
Siblings	8.5%	10.2%
Children	12.3%	13.9%
Uncles and Aunts	2.0%	3.6%
Children by 2 Schizophrenic Parents	36.6%	43.3%

One can only conclude that the evidence for a genetic etiology of schizophrenia is powerful. There is, however, the troublesome fact that many identical twins are discordant for schizophrenia, and that even two schizophrenic parents produce no more than 45% of their offspring who can be defined as schizophrenic. This poses the question as to whether this discrepancy might be due to experiential factors. Whatever the explanation may be, it does not seem to be a function of the rearing by a schizophrenic adult. The adoption studies indicate that the incidence of schizophrenia is not increased by being reared by a schizophrenic foster parent (185).

After a century of backbreaking scientific labor, we appear to have established the genetic nature of schizophrenia, although unfortunately this cannot be the complete explanation. Other factors must be included, which are not yet self-evident. However, the notion that schizophrenia is a learned behavior has been eroded as has the concept that a schizophrenic parent teaches the language of irrationality to the offspring. This does not negate Freud's analysis of Schreber, it simply places it in a different perspective. It now seems more likely that Schreber had a genetic predisposition to schizophrenia. The stresses in his life probably made his latent tendencies overt. The complex content of his delusions reflected his conflicts, but the fact that his conflicts were converted into delusions rather than neurotic concerns, was a function of his biology. A more succinct version would be that culture, contemporary experiences, and personal conflicts determine the content of delusions, but they do not explain the psychosis.

There are clear indications of a genetic defect. Once it is established, one is forced by the mandates of molecular biology to postulate that the defect is coded in DNA, the genetic material. Since DNA goes to RNA and thence to protein, one must assume that a defective, dimiminished or absent protein results. This would mean, presumably, that the protein would probably be structural or enzymatic. In the case of schizophrenia, if the protein be structural, related to the circuitry of the brain, we indeed do have a long and painful task to pursue. Hopefully the defect may prove to be enzymatic. Unfortunately there is no animal model for schizophrenia, and the brains of schizophrenics are not readily available for enzymatic analyses. Even if brains were accessible, we would have the unenviable task of deciding where to look and which enzyme to approach.

These are certainly some of the dilemmas which create an impasse.

DR. KETY: The traditional genetic techniques, so useful in many areas of medicine are not so readily applied in psychiatry be-

Study of adopted versus natural children

cause of the lack of objective indices such as blood tests, XRay findings, biopsy and autopsy information. The diagnosis of psychiatric illnesses depends on subjective judgments which though sensitive, are nonetheless corruptible. Since the tendency of mental illness to cluster in families, or the concordance of schizophrenia in monozygotic co-twins (70) may reflect both genetic and environmental influences as Dr. Bliss has just pointed out, observations are not conclusive evidence for the importance of genetic factors in that disorder.

For those reasons, three of us, David Rosenthal, Paul Wender and I, in 1962 when we were all at NIMH, decided that adoption offered a means of disentangling genetic and environmental factors in schizophrenia. To minimize ascertainment bias we decided to start with a total national sample of adults adopted in infancy, find the schizophrenics among them, and determine to what extent that disorder occurred in their genetic and environmental families.

Denmark and other Scandinavian countries maintain excellent records of legal adoptions, admissions to mental hospitals and clinics as well as comprehensive population registers. We were fortunate in obtaining the collaboration of Dr. Fini Schulsinger of the Psychiatric Department of the Kommune Hospitalet in Copenhagen, and through him, were given access to these records in Denmark, because of the authorities' interest in our research and our assurances of complete confidentiality.

Diagnostic Criteria

Beginning in 1963 we compiled a register of all individuals, now between 25 and 50 years of age, who were legally adopted by other than biological relatives through the courts of Greater Copenhagen. The sample consisted of nearly 5500 adopted individuals. The next task was to find the schizophrenics among them and before that to agree on what we would call schizophrenia. We began with the concept of schizophrenia as described by Kraepelin, as a severe, chronic, debilitating mental disorder of insidious onset, marked by bizarre thought, speech or behavior, flatness of affect, dissociation between feeling and thought, hallucinations, delusions, withdrawal from social interaction, and with marked handicaps in terms of sexual, marital, educational, occupational and social adjustments.

In addition, however, since the concept has been broadened in America, we included two other syndromes called schizophrenia in this country: latent or borderline schizophrenia and acute schizophrenic reaction. The American Psychiatric Diagnostic Manual defines latent schizophrenia as a syndrome very much like chronic schizophrenia in terms of the quality of the symptoms but differing in intensity. It is still chronic but less debilitating. There is flatness of affect and bizarre thinking along with manifestations of schizophrenia but in a more subtle form and patients are much less handicapped. The latent schizophrenic makes some kind of marginal adjustment and remains out of mental hospitals for the greater part or all of his life.

Acute schizophrenia was identified as an acute psychosis lacking the chronicity of schizophrenia and the progressive downhill course. It has in common with schizophrenia only that it is a psychosis and that during the acute psychotic episode it may resemble the exacerbated psychosis that sometimes occurs in schizophrenia.

At any rate once we all accepted the three designations and their criteria for the purposes of this study, there was considerable agreement among us.

We then searched the psychiatric registers for the 5500 names of the adoptees and we found that about 10% had some record of hospital or psychiatric clinic admission. This is what one would have expected in the population at large. Since we had to make our own diagnoses in the case of each of the adoptees with a history of psychiatric illness, the record was obtained from the institution. It was abstracted by a Danish psychiatrist, translated into English, and edited to remove any information of family history which would have prejudiced us. Then the four of us, Schulsinger in Denmark, and Rosenthal, Wender and I in the United States, read these edited abstracts. Each of us made an individual diagnosis in each case: (1) classical or what we call chronic schizophrenia; (2) acute schizophrenic reaction; (3) latent schizophrenia; (4) possible schizophrenia; or (5) no schizophrenia. Then we compared with individual assessments. When all four noted "definite schizophrenia", that person became an index case; where all four said, "not schizophrenic" that individual was rejected; where one or more suggested schizophrenia, more information was obtained and we held a conference. If we could all agree on a diagnosis of schizophrenia by the end of that conference, that became an index case. In that way we agreed on 33 schizophrenic index cases out of the 5500 adoptees in Copenhagen, a .6% incidence, but since all subjects had not lived through the age of risk, that is not out of line with an 0.8% lifetime incidence expected in the general population. As we were using a broader definition for the term schizophrenia than the Danes ordinarily use, it may be that we have a lower incidence of classical schizophrenia in this adopted sample than one would have expected.

To satisfy ourselves on this point we obtained a sample (5483 persons) from the non-adopted population living in Greater Copenhagen matched for age, sex, socioeconomic class of the rearing family with the 5483 individuals in the adopted sample. Applying our criteria to cases of mental illness found by a search of the Psychiatric Register revealed almost exactly the same incidence of schizophrenia as defined here in the adopted as in the non-adopted sample.

Continuing our research, the 33 index cases of adopted persons in whom the diagnosis of schizophrenia was agreed upon were matched as to age, sex, socioeconomic class of the rearing family and length of time spent with the biological mother with 74 adopted control cases who had never been seen in a psychiatric facility for mental illness.

We shuffled the index and control probands' names together and sent them back to Schulsinger, with a request that his staff identify the relatives of these probands through the adoption records and their "folkeregister" in which the names of the other children the parents have had would be recorded. In that way we identified the biological and adoptive parents, siblings and half-siblings of the schizophrenic and control probands, a total of approximately 500 relatives. We did not feel that the adoptees were old enough to have had many children of their own and the children in any case would not have lived long enough to develop schizophrenia, so there was no point in identifying their children.

Of the four parents of each adoptee, the biological father is the only one that would be in any doubt, although this is probably less doubtful in Denmark than it would be in most countries. In Denmark it is not enough for a woman to identify the father of a child that she is giving up for adoption. The putative father has to acknowledge paternity and must contribute to the support of maternity care and to the legal cost of adoption, so it is likely that he thought he was the father. Studies in the United States using blood group characteristics indicate than an error of as much as 20% could occur in acknowledged paternity. We feel the error may be as high as that but is likely to be considerably lower in Denmark.

We recognized recently that the names of the siblings and half-siblings recorded in the folkeregister would probably not be complete for other children of the biological parents born out of wedlock. Therefore we may have missed some half-siblings whom we hope eventually to recover among the 14,500 adoptees we have now identified in all of Denmark.

Variants of Schizophrenia

We were interested in any kind of mental illness, however mild, that occurred in those 500 biological and adoptive relatives. In addition to the Psychiatric Register a search was made of military and court records which may have contained psychiatric information on these relatives. Information we obtained on any relative was abstracted, translated into English, and edited to remove any clues that a sophisticated reader could use to guess that this was a biological or an adoptive relative of an index or control proband. These abstracted records were sent to each of us to read and diagnose independently and subsequently in consensus. Since we have two families for these adopted schizophrenics we can now test the importance of genetic and environmental factors. If the familial incidence of schizophrenia is entirely the result of genetic influences, we should find only the genetic family to have the high incidence of schizophrenia. If it is entirely environmental we should find the high incidence in only the adoptive family. If both factors were necessary we might find schizophrenia in both families or in neither.

Those who have studied the relatives of schizophrenics have been impressed with the finding that some of them are classical schizophrenics but a substantial number have many of the features of schizophrenia although that diagnosis would be inappropriate for them. This has resulted in the concept called "schizoidia" by some to designate a vaguely defined group with considerably milder symptoms, but which may be related to schizophrenia. In our selection of the index cases we were sometimes confronted by individuals who fell into that category but whom we excluded since we felt that our index cases should be representative of definite schizophrenia. In characterizing their relatives, however, we could afford no such luxury of excluding some where our diagnosis was not definite. We permitted a diagnosis of "uncertain schizophrenia" for those individuals where schizophrenia appeared to be the best diagnosis although some of the features were lacking or atypical, making the diagnosis unsure. There were some individuals with even milder characteristics who would fit the diagnoses listed in the American Diagnostic Manual as "schizoid personality" or "inadequate personality". We decided to use all of these diagnoses wherever they seemed most appropriate in reviewing the relatives with the thought that this would give us the opportunity of testing the genetic relatedness of these categories to schizophrenia. We thus had a spectrum of diagnoses of decreasing severity beginning with chronic schizophrenia (the syndrome described by Kraepelin) and including the two forms of schizophrenia recognized in America - latent or borderline schizophrenia and acute schizophrenic reaction, then going through uncertain schizophrenia, schizoid personality, and finally - inadequate personality. This we called the schizophrenic spectrum and in the course of our later studies have tested the various components for their relationship to schizophrenia.

What we found in our first study, published in 1968 (105), in a sub-group of 33 index cases and their controls who had been adopted through the courts of Greater Copenhagen, was a statistically significant increase in schizophrenia spectrum diagnoses in the biological relatives of the schizophrenic adoptees as compared with their controls. The adoptive relatives, with whom the schizophrenic adoptees had been reared, showed no greater prevalence of schizophrenia spectrum disorders than did the adoptive relatives of controls. The number of spectrum diagnoses, however, in the relatives was so small that the individual diagnostic categories could not be tested for significance.

Genetic versus Environmental Factors

Furthermore, although we had presumptive evidence that genetic factors operated significantly in the transmission of schizophrenia-related disorders, we had little information on the operation of environmental factors. What we could say was that the presence of schizophrenia in the rearing family was not necessary for its development in the adoptee, but we had no information on possible schizophrenogenic personality, childrearing practices and other environmental influences of possible relevance. For these reasons we decided to extend the original study in two other directions.

The first extension was to continue the compilation of adoptees for all of Denmark. This has been completed and the total adoptee sample for all of Denmark is approximately 14,500 in number, of whom 74 have been agreed upon as index cases with a definite diagnosis of schizophrenia. An equal number of matched control adoptees have also been selected, and approximately 1100 biological and adoptive relatives of the 148 probands have been identified through the population registers. Information on mental illness in these relatives has been obtained, abstracted, translated into English and edited. Blind, independent and consensus diagnoses have been made. The results of this extension which will be published at a later date support and amplify the initial study. There is a highly significant concentration of schizophrenic spectrum diagnoses and also diagnoses of definite schizophrenia in the biological relatives of the schizophrenic adoptees compared to their controls and no difference with regard to these diagnoses in the adoptive relatives. Furthermore, most of the diagnoses of definite schizophrenia in the relatives are found in the biological families of the index cases who received a diagnosis of chronic, classical schizophrenia. The adoptees diagnosed as borderline or latent schizophrenia showed a somewhat elevated prevalence of definite schizophrenia in their biological relatives whereas those diagnosed as acute schizophrenic reaction did not. This permits the tentative inference which will have to be further explored that borderline schizophrenia may be a valid subgroup of schizophrenic illness whereas what is called acute schizophrenic reaction may not be.

The second extension was a more intensive examination of the relatives of the 33 index adoptees and their controls in the Greater Copenhagen sample. We felt that psychiatric interviews on as many of these relatives as would cooperate would yield considerably more information than we had hitherto obtained regarding mental illness and personality traits in them and that some evaluation would be possible of a large number of environmental variables in the interviews with adoptive relatives (66). We were fortunate to obtain the collaboration of Dr. Bjørn Jacobsen, who spent two years conducting these interviews.

Deaths including Suicide among Schizophrenics

Of the 512 relatives who had been identified, 119 had died by the time of the completion of this study in 1973. The deaths were not randomly distributed among the relatives but showed a significant increase in the biological index relatives (35/173 vs. 13/174, $p = .001$). The death certificates have been examined in these individuals and we find that all of the excess in deaths can be accounted for on the basis of suicide, presumptive suicide, and a few accidental or violent deaths. The death rate from natural causes is not significantly higher in the biological index relatives than in their controls.

DR. BIGELOW: Does this high incidence of suicide suggest that the diagnosis might be depression and not schizophrenia?

DR. KETY: It might be depression, but the incidence of suicide among schizophrenics is almost as great as among depressed people.

DR. CANCRO: Schizophrenics are more successful in their attempts. They do it with fewer efforts.

DR. BLISS: There have been a number of reports on this. The incidence of suicide is not as high as it is in depression but it is very high.

DR. KETY: We have, however, probably lost a number of schizophrenic relatives among the suicides.

Psychiatric Patterns Among Relatives

An additional 26 relatives had emigrated beyond Scandinavia. Of those who were alive and residing in Denmark, Norway or Sweden, Dr. Jacobsen achieved the cooperation of 90% in an exhaustive interview which usually took two hours and sometimes considerably longer. An additional 5% who refused to participate did so in a conversation with Jacobsen that was sufficiently productive to yield a fair amount of information. The 23 relatives who refused interview and gave no information were randomly distributed with no significant concentration in any particular group. It is clear that Dr. Jacobsen achieved a remarkably high degree of cooperation and representativeness in his interviews.

The content of these interviews was carefully thought out. They included a wealth of information about the life experience of the individual: educational, occupational, sexual, marital, psychological histories, peer relationships, kind of home, parental personality, relationship with and between parents, symptoms of psychopathology as a child, adolescent and at present, and a complete mental status examination. The interviews were conducted in Danish naturally but Dr. Jacobsen dictated a complete summary in English. The transcript of each interview ran 35 pages, half of which were checklists, the other half narrative material. These transcripts were then edited to remove any clues that would provide a sophisticated reader with hints that the subject was a biological or adoptive relative of an index or control adoptee. Dr. Jacobsen did not know at the time he began the interview the relationship of the subject to a proband and avoided asking questions pertinent thereto. If the subject volunteered information that might have provided a clue Dr. Jacobsen noted this in a designated place on the interview as well as any other hunches he had derived regarding the relationship. All of that was deleted in the editing process but at the end of the study was evaluated and indicated that his correct hunches were relatively few in number and would not have influenced the results. Since this was a Copenhagen sample Dr. Schulsinger was ruled out as a rater since he might know some of the subjects. Instead the edited interviews were read by Rosenthal, Wender and me who did not know any of the subjects or their relationships. Each of us made a diagnosis and then a consensus diagnosis was hammered out. Table XII indicates the diagnoses outside of the schizophrenia spectrum that were made in the various groups of relatives. Nearly 40% of the subjects were judged to have no psychiatric diagnosis while 50% received a diagnosis outside of the schizophrenia spectrum. The distribution, however, was quite constant in the four groups. There was no clustering of organic, neurotic, affective or personality disorders among those who were genetically related to the schizophrenic adoptees, and thus no evidence in this study of a genetic relationship of schizophrenia with other mental disorders.

When we turn to the schizophrenia spectrum however, (Table XIII) the situation is quite different and there is a highly significant concentration of these disorders in the biological relatives of index cases in comparison with their controls. This is also true for our diagnosis of definite schizophrenia and for the most severe of the components, chronic schizophrenia. The same significant concentration of definite and chronic schizophrenia is found in the biological relatives on the basis of the independent diagnoses of each of the three raters. What we have called "uncertain schizophrenia" individually and in consensus is also very significantly concentrated only in the biological relatives of the schizophrenic index cases as compared with

TABLE XII

CONSENSUS DIAGNOSES OUTSIDE THE SCHIZOPHRENIA SPECTRUM IN BIOLOGICAL AND ADOPTIVE RELATIVES

	Interviewed Relatives, in %			
	Index Group		Control Group	
Diagnosis	Biological (N=81)	Adoptive (N=31)	Biological (N=121)	Adoptive (N=41)
Psychiatrically normal	37	36	41	27
Organic illness	9	16	5	15
Neurotic illness	5	10	5	5
Affective illness	2	3	9	7
Personality disorder	33	26	32	37
All nonschizophrenic diagnoses	49	55	51	63

* The Ns represent interviewed relatives whose consensus diagnoses were outside the schizophrenia spectrum.

TABLE XIII

PREVALENCE OF SCHIZOPHRENIA SPECTRUM DISORDERS IN THE BIOLOGICAL AND ADOPTIVE RELATIVES OF SCHIZOPHRENIC INDEX AND CONTROL SUBJECTS, FROM CONSENSUS DIAGNOSIS ON INTERVIEW

Type of Relative	Number of Relatives Identified	Diagnosis of Schizophrenia in Relatives							
		Chronic		Latent		Uncertain		Total	
		Number	%	Number	%	Number	%	Number	%
Biological relatives of schizophrenic adoptees	173	5*	2.9	6	3.5	13**	7.5	24***	13.9
Biological relatives of control adoptees	174	0	-	3	1.7	3	1.7	6	3.4
Adoptive relatives of schizophrenic adoptees	74	1	1.4	0	-	1	1.4	2	2.7
Adoptive relatives of control adoptees	91	1	1.1	1	1.1	3	3.3	5	5.5

* $p < .05$ (significances apply to differences between biological relatives of schizophrenic adoptees and controls; other differences were not significant).

** $p < .01$.

*** $p < .001$.

their controls. This warrants further efforts at characterizing this rather vague category which appears so clearly to be related to schizophrenia and this we are attempting to do. As in the case of the previous studies the adoptive relatives of the schizophrenic probands do not show an elevated prevalence of any of these categories of schizophrenia in comparison with their controls. The last component of the schizophrenia spectrum - inadequate and schizoid personality - is equally represented in the biological index relatives and their controls, so that this study does not yield evidence relating these personality disorders to schizophrenia. We have evidence, from other studies, that there may be a valid member of the schizophrenia spectrum included in a larger group of such personality disorders and we are making efforts to determine what features may distinguish the two.

Possible associated Environmental Factors

The evidence thus far presented is compatible with a genetic transmission in schizophrenia, but is not entirely conclusive, since there are possible environmental factors such as in utero influences, birth trauma, and early mothering experiences which have not been ruled out. The ideal adoption study would involve separating the child from the biological mother at the time of conception! However, there are 127 biological paternal half-siblings of index cases and controls among these relatives who can help to settle that question, since the biological paternal half-siblings did not share the same mother, neonatal mothering experience or post-natal environment with their adopted half-siblings. The only thing they shared was the same father and a certain amount of genetic overlap. The number of paternal half-siblings is almost identical for index cases and controls, but the number of those who were diagnosed as having definite or uncertain schizophrenia is markedly different, with 14 among the half-siblings of index cases and only two among the half siblings of controls ($p = .001$). If we restrict the diagnosis to definite schizophrenia the same concentration exists (8 vs. 1, $p = .015$). We regard this as compelling evidence that genetic factors operate significantly in the transmission of schizophrenia.

These data do not permit the conclusion that schizophrenia is a unitary disorder, since they are equally compatible with a syndrome of multiple etiologies and different modes of genetic transmission. Our results depend very heavily upon the relatively high incidence of schizophrenic diagnoses in the biological relatives of the index cases with classical chronic schizophrenia. Furthermore, only half of the schizophrenic adoptees had any diagnosis of definite or uncertain schizophrenia in their biological relatives. This would be compatible with the possibility that there are at least two forms of schizophrenia, one with a strong genetic basis which is weak or absent in the other.

Schizophrenia is analogous to diabetes mellitus in many ways. Both are symptom clusters or syndromes, one described by somatic and biochemical abnormalities, the other thus far by psychological. Each may have many etiologies and shows a range of intensity from severe and debilitating to latent or borderline. There is also evidence that genetic and environmental influences operate in the development of both. The medical model seems to be quite as appropriate for the one as for the other.

DR. ALTSCHULE: How well did your raters agree on the diagnosis?

DR. KETY: In arriving at a diagnosis Wender, Rosenthal and I agreed very well on classical schizophrenia. If one of us independently called a person chronic schizophrenia, the others invariably made the diagnosis of definite schizophrenia, usually chronic, in one or two instances the others differed in terms of intensity. The others may have called it borderline instead of chronic, but in no case where one of us said chronic schizophrenia, did the others fail to call it definite schizophrenia. In borderline schizophrenia our inter-rater reliability was not as good. We had about 80% agreement on the diagnosis of borderline schizophrenia and somewhat less on the diagnosis of uncertain schizophrenia, which is about what psychiatrists in general do. We think that these other categories, what we call "borderline schizophrenia" and "uncertain schizophrenia", require a great deal of further work in order to define them more closely.

DR. WERTHESSEN: I would think that you would get more solid proof that you are on the right track by recalculating the data on the basis of what each one of the three of you said independently.

DR. KETY: As a matter of fact we have done that. We have analyzed the data on the basis of each of our individual diagnoses. Each one of us independently found the same conclusion as we found in the consensus, i.e. a statistically significant contribution of definite schizophrenia and of "uncertain schizophrenia" in the biological relatives of the schizophrenic adoptees. In the case of "schizoid" or "inadequate personality" only one of us found a significant increase there; the other two and our consensus found equal numbers in the index and control relatives.

DR. WERTHESSEN: But you see this is precisely the same problem as when you are looking at a population of cells through a microscope. Some people can see epithelial versus cornified cells, dead or live sperm. So that if you have two or three people look at the same slide and run your tables through the same group of three you come out with the same scientific conclusion, independent of the way they split their differences, then I think you have established an unattackable case.

DR. BLISS: Would you comment on a recent study of infants adopted by foster parents who have had schizophrenia? One would expect a higher incidence of schizophrenia in these adopted children if the environmental hypothesis was correct.

Cross Fostering Study

DR. KETY: That is the Cross Fostering Study of Wender and the rest of us (186). Cross fostering with animals is done when offspring are reared with an affected parent and other offspring are reared with a normal parent. You then separate the offspring in terms of their genetic predisposition. We could only approximate that with human beings. We were able to review the biological parents, and pick those on whom we could make a diagnosis of schizophrenia. Their biological children would be the children of a schizophrenic parent. As opposed to Heston's Study where everybody knew that these were the children of schizophrenics, in our study practically no one knew they were children of schizophrenics because the schizophrenia occurred in the biological relatives on the average twelve years after the child was adopted and there was little, if any, transfer of information. Then we reviewed the adoptive relatives and found those where a diagnosis of schizophrenia or psychosis could be made, and in this way we now have four populations:

1. The biological children of schizophrenics adopted and reared by normals;

2. The children of one or two schizophrenic parents reared by these parents without adoption;

3. The biological children of normals adopted and reared by a schizophrenic or otherwise psychotic parent;

4. The biological children of normals adopted and reared by normals.

The two important comparisons are between the biological children of schizophrenics reared by normals and the biological children of normals reared by schizophrenics. The findings were consistent with genetic transmission; only the people genetically related to schizophrenics developed schizophrenia, regardless of the presence or absence of psychosis in the rearing family.

DR. CANCRO: Do you have enough information about the biologic parents, who gave up their children for adoption, distinguishing between those who became pregnant after schizophrenic episodes as opposed to those who had the pregnancy before they had their first schizophrenic episode?

It would then be possible to look at the children who were put up for adoption when the conception took place prior to the first psychotic episode as opposed to those where the conception took place afterwards?

DR. KETY: Only a very few of the biological parents were known to be schizophrenic at the time the child was adopted. We may be losing the most severe cases in this sample perhaps because they are never born. Denmark has a very enlightened abortion law so that a mother may obtain a legal abortion if there is a history of mental illness in her or the father's family. Furthermore the children of an overt schizophrenic are usually not adopted but may be reared in institutions so that they do not appear in the list of adoptees.

DR. WOLF: This makes it all the more extraordinary that your figures for prevalence match those in other parts of the country where there are not these skimmers.

DR. KETY: Yes, but our diagnosis of schizophrenia may be broader than that usually used.

DR. CANCRO: Right. That broadening may be also compensated for by narrowing effects.

DR. KETY: There are other factors that should be raised. What about assortative mating or assortative placement? We do not know about assortative mating yet, although we may assume that it occurs to some extent. But assortative placement. Is there any? We have found little evidence of selective placement. It is hard to see how either of these factors would affect our results. You would have to assume that the adoption agency or the social worker or the judge could identify a schizophrenogenic home in which to rear particular children.

DR. CANCRO: He could not do it, even if he wanted to.

DR. KETY: Another question has been raised and this is a reasonable question. Is it possible that we are looking at a very special kind of schizophrenic index case because these are people who developed schizophrenia in spite of having a randomized, normal environment? Therefore, these people may have been very heavily genetically loaded, and one may say that they would have become schizophrenic under any circumstance. Our results would then apply to at least one kind --

but not necessarily all kinds -- of schizophrenia. However, the incidence of schizophrenia in our adopted sample is almost exactly the same as that in an unadopted sample we have compiled in Copenhagen and carefully matched for age, sex and socioeconomic status of the rearing family. If environmental factors were of leading importance in schizophrenia, the incidence of schizophrenia in an adopted population should be significantly lower than the incidence in a non-adopted population. And yet the importance of an environmental factor is suggested by the 50% discordance in the twin studies. Thus genetic factors cannot account for everything. It could be that the environmental factor does not segregate by families.

DR. CANCRO: Well, I think there are two explanations that are not really too forced. One argument held by some is that these children are different. They relate differently to their parents and they tend to create a microenvironment in virtually any home in which you place them, which is as likely to occur in an adoptive as in a nonadoptive family. This explanation covers the data without being totally spiritualistic, but you feel the strain when you listen to the argument. I think one can make a more cogent case for the other explanation. Yet you have to be very careful to stop using environment so broadly as to include everything from the concentration of extracellular electrolytes to whether the mother is depressed or not. If we use the term environment in this very broad sense of non-genetics, we can argue that experiences likely to precipitate a psychosis are relatively ubiquitous in a given culture and that these stresses are as likely to appear in the adoptive as in the biological home. Unless we make a distinction between vulnerability and the precipitating stresses leading to a psychotic decompensation, we get into some sticky situations. I think it is a useful distinction if you think of vulnerability as something that is probably prenatal and that the precipitating stresses are probably related to the social process of maturation. Therefore if you have vulnerability and the right developmental stresses, classical schizophrenia is the form the illness is likely to take.

DR. KETY: I agree with your latter suggestion; an environmental factor that hit individuals rather than families but was still rather widespread and capable of producing schizophrenia in genetically vulnerable individuals would account for our data and the discordance in monozygotic twins. From these interviews we have been trying to ascertain what environmental factors operate. We have been analyzing the interviews which contain a great deal of experiential information and we have just run an analysis of 800 variables which include everything under the sun relating to personality of parents, type of home, and also for characteristics of the individual. We have done this for the biological siblings and adopted siblings and half siblings, the probands' mothers, fathers and parents together, and we have compared the index relatives with the control relatives on the biological side and on the adopted side. Now we recognize that when many hypotheses are run simultaneously one will come up with significant values merely by chance: 5% of the variables will have a p value of .05 and 1% will have a p value of .01.

What we have found so far is interesting. In the 40 adoptive relatives we find no more differences than can be accounted for by chance in any of these 800 variables, and the differences that we find do not make too much sense. On the other hand in the biological relatives we find three times as many differences as can be accounted for by chance which suggests that some of the differences are real. However, the differences we find have to do with the symptomatology of schizophrenia. So we may merely be confirming the fact that we have schizophrenics in the biological relatives. But, the number of cases is small; many of the adoptive relatives are dead and there are very few siblings. One could argue that we have still not given the environmental variables a complete rundown. Certainly we have not given them the kind of rundown that Lidz gives them by looking intensively into the family relationships, but at least we have done it blindly.

DR. WOLF: Can you say anything about handedness in schizophrenia?

DR. BIGELOW: There is nothing significant there.

DR. WYATT: There is no preponderance of left-handed schizophrenics?

DR. CANCRO: No, but there is in epilepsy.

Laterality in Identical Twins

DR. WYATT: I found out something relevant to handedness just yesterday. I have two-year-old fraternal twin boys. We were riding on a merry-go-round. When you have twins you tend to notice other twins. In front of us was a set of identical twins. One was sucking his right thumb and the other his left. I was astonished to see that because I thought that sucking thumbs was very indicative of handedness -- opposite to the dominant hand. I expressed my thoughts to some people and they said, "Well, of course, there is mirror imaging." Every teacher looks for the right and left handed twin, if they are identical. So I quickly ran through the literature and found that about 20% of identical twins are opposite handed. Then I looked at Bill Pollin's collection of monozygotic twins who were discordant for schizophrenia (147). There were laterality tests on ten pairs and it turns out that there were many reversals in handedness, eyedness and footedness. That means where one twin was right-handed, the other twin was lefthanded; where one twin was righteyed, the other twin was lefteyed, etc. Presumably dominance is, in part, genetically determined. I do not, however, have the evidence to support this point. In this set of subjects, there were 13 reversals out of the 27 factors that I looked at.

DR. KETY: Was there more reversal in the twins discordant or those concordant for schizophrenia?

DR. WYATT: That I do not know. I only examined the discordant twins. The interesting thing here is that you have something that is presumably in part genetically determined which seems almost to be random, in this group at least. You have real reversals within a genetically determined factor. The point is that identical twins are not identical.

DR. WOLF: But this should be testable in the population of twins that are discordant for schizophrenia. If some of those are going to have one member of the pair righthanded and one member of the pair lefthanded, it should be possible to state whether or not schizophrenia was more common among the righthanded or lefthanded.

DR. COHEN: In developmental systems, different cytoplasmic polarities may be established in the egg by virtue of the position of the zygotes, even if they have the same DNA.

DR. WOLF: Skeletal abnormalities are also often in mirror arrangement, that is, if one twin has a dentigerous cyst of the right side, the identical twin is more likely to have dentigerous cyst on the left than he is on the right. The same is true of other skeletal abnormalities.

DR. WYATT: Right, but it occurs very early on in development.

DR. LIPTON: Dr. Boklage, who recently arrived at our University, thinks that laterality has a good deal to do with schizophrenia. He has noted that in almost all of the discordant monozygotic twins studied by Gottesman and Shields, one of the twins is lefthanded. Why this should be is unclear, but laterality is set up early in embryo genesis and Boklage has some interesting speculations about laterality in brain function as related to schizophrenia (12).

The Possible Contribution of Parental Behavior

Gene, may I come back to the point you made earlier about learning to be schizophrenic. If you are thinking of learning to be schizophrenic, like learning French, the way to learn French is to be raised by French parents, and the analogy that you are using is that the way to learn to be schizophrenic is to be raised by schizophrenic parents. It may well be, however, that it is quite different and that a schizophrenic parent is not necessarily a schizophrenogenic parent. It may be that the schizophrenogenic parent is not schizophrenic but has other traits, neurotic or otherwise which may damage the child. I would not give up the concept that parental influences may have such determining roles because they do in mental retardation and in emotional disturbances among animals. By the same token the mere fact that the incidence of schizophrenia was found to be as high among children raised by nonschizophrenic parents as by schizophrenic parents does not rule out parental influences as etiologic factors. I would give up the concept that early psychological life experiences have nothing to do with schizophrenia, only very, very reluctantly, and after only the most cogent evidence.

DR. BLISS: I have not abandoned the notion that some schizophrenias may be learned behaviors. However, it is useful to postulate that such is not the case. Then one can ask what biological influences other than genetic factors may be critical. It then forces one to consider more seriously such possibilities as brain damage.

DR. LIPTON: Well it is worth it but what I am saying has its somatic implications. If I may go back to the analogy of mental retardation. I was recently reviewing this and probably your evidence will provide some nice studies on this. The question has been raised by Cravioto of why is it that in an impoverished Guatemalan village, where all of the children are mildly malnourished, only about 3 to 5% of them will develop Kwashiorkor. In an epidemiological study, Cravioto attempts to relate this to how many children in the family, or the level of education of the mother, or the quality of sanitary facilities and he finds no relationship at all. But he finds an interesting relationship in what he calls the microenvironment. Cravioto is a nutritionist not a psychiatrist. Yet he says that the incidence of Kwashiorkor correlates with what he calls the microenvironment, meaning how much does the mother play with or talk to her child, or how many toys does the child have. When these are deficient, then the child is more likely to develop Kwashiorkor than when otherwise. Thus the issue is what kind of parents the child has; what kind of emotional or cognitive environment do they offer the child? It seems to me a parallel might very well exist with respect to schizophrenia.

DR. KETY: You are assuming that it is a psychological thing that is influencing the Kwashiorkor, but a family that talks to the child and gives it more toys, and so on, could also be giving it more protein.

DR. LIPTON: No. He examined that. The study was conducted in a marginal geographic area, where all the children are malnourished for protein.

DR. KETY: He knows in general the protein intake of the region or even of the family, but he cannot examine the protein intake of the child. Within the family where there is a great concern for the child, the parents could just be giving the child better food and depriving themselves.

DR. LIPTON: That may well be, and giving a child extra food on demand is a form of paying attention or stimulating the child also. Furthermore the child becomes expectant apparently because of the attention that is paid. He will demand more food whereas the other child who is ignored may very well say, "What the hell, you're not feeding me when I cry for it, so I'll stop demanding. I'll learn to do without." Such a child may withdraw interpersonally and initiate a circular process in which he expects little, demands little and gets little. This could lead to Kwashiorkor. The point I want to emphasize here is that it is quite possible, and there are some animal data to support this, that the healthy brain can withstand many types of environmental vicissitudes. On the other hand the damaged brain may be much more susceptible to detrimental environments. It may not matter whether that brain is damaged genetically, by birth, injury or malnutrition, by endocrinopathy or by toxic substances in the physical chemical environment. But when it is damaged, genetically or otherwise, I would venture to guess that behavior is uniquely susceptible to the parental or psychosocial environment in which the organism is reared. I wonder whether the groups studying the Denmark children are focusing on this type of interactive effect.

Brain Injury in Genetically Predisposed Individuals

DR. WOLF: I wondered if Dr. Kety might say something about the work of Sarnoff Mednick? Dr. Berle and I were discussing reports of genetically predisposed individuals who had been exposed to some sort of brain insult, hypoxia during the birth process for example, who later developed behavioral disturbances of a schizophrenic type. Mednick noted a disturbance in the extinction and habituation mechanisms in these individuals and had some reason to suspect a disorder in hippocampal function. Would this imply a disturbance in dopamine or dopamine receptor function in the hippocampal area?

DR. KETY: Mednick and Schulsinger were the first to press the high risk approach to schizophrenia although it had been suggested earlier. In the course of longitudinal studies on children of schizophrenics compared to children of controls, Mednick found certain changes in autonomic response which appeared to characterize a large segment of the high risk children (124). These are children who would be expected to have a 10% breakdown into schizophrenia at some time. At the present time that finding is somewhat controversial. Van Dyke has attempted to replicate it and failed to find the distinguishing autonomic changes Mednick reported. Neither has the Erlenmeyer-Kimling group been able to substantiate the findings. But it is an interesting finding, and it is still being examined. Mednick also hypothesized that birth injury might account for one of the environmental factors in schizophrenia. In reviewing the literature, he did focus on the hippocampus as being a likely site of hypoxia or birth trauma. In Denmark where the birth records are very complete they have been reviewing a series of 9000 births separating the traumatic from the non-traumatic (105). These births took place long enough ago so that the children are now adults. They have been trying to see whether there is a higher incidence of schizophrenia in those where there was birth trauma than in those where there was not. To the best of my knowledge that work has not yet been completed. So far I have not been impressed with the evidence that birth trauma plays a crucial role but Dr. Bliss, who knows more than I do on this subject, made reference to some studies suggesting that it might be significant.

DR. BLISS: It remains controversial. The findings are contradictory. There are some papers which report a higher incidence of complications of pregnancy in schizophrenic mothers; others do not. There are reports of subtle neurological changes in some of the offspring of schizophrenic mothers. There are enough reports to make one suspicious and perhaps to draw the tentative conclusion that some offspring of schizophrenics may very well have brain damage. Certain studies also pose the intriguing possibility that some schizophrenias may be diseases with a neuropathy that has never been detected or described.

DR. CANCRO: I would like to ask one methodological question. Are these perinatal and prenatal data gathered retrospectively or prospectively? If they are done retrospectively the need for very special control groups becomes critical. Those of us who have been interested in longitudinal research and try to engage in it, find that retrospective histories, particularly from families where there is guilt or pain associated with the illness of a child, become like a Rorschach. There is a lot of room for projection and for recalling all kinds of things that otherwise would not have been recalled and may not have happened. The ideal study would be prospective, one in which assessments as to the presence or absence of pre-and perinatal injuries are made during pregnancy and at the time of birth.

DR. KETY: That was the excellent feature of the Mednick Schulsinger study. It was prospective. I would like to mention one thing à propos the brain damage hypothesis. Goldfarb, who studied schizophrenic children for many years, found signs of organic features in some of these children as opposed to other children who seemed to have functional schizophrenia with no signs of neurologic deficit (66). The interesting thing he found was that in those who had organic deficit, the parents were much more intact than in those who did not, suggesting that there were two kinds of schizophrenia, one with a greater genetic load and another with a more traumatic overlay.

Paul Wender is going over the index cases that I talked about. (106). Remember that half of them had a strong genetic load in their biological relatives and half of them did not. He is going over these index cases "blind" to see if he can separate them into cases typical or atypical of functional schizophrenia. The atypical are defined as those who seem to have histories of encephalitis, brain trauma, or organic features. He will try to see whether the clinical features correlate in any way with the high or low genetic load.

The "runt" of the twins

DR. SNYDER: In terms of trying to determine organicity in schizophrenia, I was wondering if in the identical twins discordant for schizophrenia where the schizophrenic twin was said to be sicker at birth or weighed less, was this difference documented? Did they have the objective records or did the investigator rely on the retrospective reports of the parents?

DR. KETY: Pollin (147) found remarkable consistency in his discordant twins with the schizophrenic twin being lower in birth weight, weaker, and so on, although the data have not been substantiated in all instances. His seemed to be an unusual sample. Some data were retrospective but some were collected from hospital records of the births of the twins. But à propos of the question, to what extent is discordance in monozygotic twins accounted for by seeing phenocopies of schizophrenia as opposed to true schizophrenia, there is the study by Margit Fisher of the children of discordant monozygotic twins, referred to by Dr. Bliss (49). The discordant monozygotic twins had the same incidence of schizophrenia in their children, whether or not they themselves were schizophrenic, which seemed to rule out, in that sample at least, a non-genetic explanation for the discordance.

DR. BLISS: You have discordant twins, one having schizophrenia and the other not, but presumably they both have the same genetic loading.

DR. KETY: Yes, the Fisher study seems to rule out the phenotypic kind of schizophrenia in the discordant pair, but we must remember that only a few of the twin pairs had offspring who became schizophrenic.

DR. WERTHESSEN: Does it really? If you have a multi-gene basis for your susceptibility?

DR. KETY: The only thing it rules out is the possibility that many of the discordant HZ twin index probands were suffering from a different kind of schizophrenia, one for which there was not a strong genetic basis. That seems to be ruled out, because the discordant twin is still capable of transmitting to his children the same genetic risks.

DR. WERTHESSEN: The fact that one monozygotic twin became schizophrenic implies either (a) that a nongenetic, i.e., environmental or other, factor occurred in his life that permitted expression of the genetic tendency to schizophrenia in that twin; or (b) that such a factor was not operating in the life of the twin that remained normal; or (c) that if it did operate, the slight but real difference between such twins permitted the difference in response.

It also seems highly probable to me that susceptibility to develop schizophrenia cannot be related to a single gene. I would prefer to view the situation as far more complex than the genetics of hair color, or even carbohydrate intolerance or frank diabetes. A number of specific genes plus modifiers of their expression seems more realistic to me.

I would tend to assume that several kinds of schizophrenia with varying levels of severity and ease of expression of the genetic tendency in response to an external stimulus will eventually be worked out. Under those circumstances, "concordant twins" would most probably be twins whose genetic tendency is easily expressed. The "discordant twins" probably represent a genetic situation requiring additionally a severe "stress."

The implications of the discordant twins' offspring showing the same level of susceptibility are to me quite disturbing. It seems proper to assume that the twins did not share their spouse. Thus, if two different sets of genes mated, in effect, to a single individual produced the same frequency of an aberrant state, we are obliged to assume that the genes of the population are "loaded" with the potential of developing the schizophrenic state.

DR. BLISS: You make a cogent point. If you are going to do first rate genetic studies, it is important to know something about both parents and not just one.

CHAPTER V - Pathophysiology - Biochemical Considerations

Participants: Drs. Altschule, Bigelow, Bliss, Cancro, Cohen, Gjessing, Haber, Kety, Lipton, Snyder, Wolf, Wyatt.

DR. SNYDER: One aspect of brain biochemistry that applies to the study of mental illness has to do with neurotransmitters. All information processing in the brain requires the participation of neurotransmitters in the regulation of synaptic activity. Moreover the actions of the psychotropic drugs depend for their effects on neurotransmitter function.

Action of Phenothiazines

Data relating the neurotransmitters to schizophrenia are at best indirect. That is, although there is no demonstrated neurotransmitter excess or deficiency in the brains of schizophrenics we know of drugs acting on neurotransmitters that make schizophrenics better and others that make them worse. Thus, given a drug with a selective action in a mental disease and knowing how that drug acts in the brain, one can infer indirectly that the biochemical mechanism of action of that drug is related in some way to the pathophysiology of the disease and its symptoms. Phenothiazines block dopamine receptors roughly in proportion to their therapeutic efficacy. Amphetamines and related drugs, on the other hand, exacerbate schizophrenic symptoms and produce schizophrenia-like psychoses. Thus they do the exact opposite of phenothiazines. They especially enhance dopamine action. A recently published report from Sweden indicated that the therapeutic effects of phenothiazines were enhanced by the addition of an agent that blocked catecholamine synthesis. Phenothiazines, of course, do not cure schizophrenia. They appear to interfere, somehow, with fundamental schizophrenic mechanisms, since they are far more effective therapeutically than barbiturate sedatives or anti-anxiety drugs such as Librium or Valium.

Amphetamines

Amphetamines aggravate the symptoms of schizophrenia. Amphetamine facilitates the actions of catecholamines such as norepinephrine and dopamine. These compounds given in small doses make schizophrenic symptoms very much worse. "Speed freaks" who use lots of amphetamines almost invariably at one time or another in their career have an acute psychosis that is often clinically indistinguishable from acute paranoid schizophrenia. Reactions to cocaine intoxication are very much like those of amphetamine, including the euphoric stimulant manifestations. Cocaine psychosis is indistinguishable from amphetamine psychosis. Both drugs act similarly by facilitating catecholamines, though by somewhat different mechanisms.

If schizophrenic symptoms are made worse by treating with amphetamines and amphetamines are acting via catecholamines, especially dopamine, the administration of the amino acid precursor of dopamine, L-dopa, used in the treatment of Parkinson's Disease should worsen schizophrenic symptoms, and indeed it does.

Anti-depressant Agents

In a number of instances a psychosis very much like the amphetamine psychosis has been observed as one of the side effects of tricyclic anti-depressant therapy. Basically the same kind of actions occur. There are two major actions of amphetamine. It releases catecholamines directly into synaptic neurons and inhibits the re-uptake inactivation. Catecholamines and probably most neurotransmitters, after they act on the synapse, are inactivated by a pump which takes them back into the nerve ending that releases them. All three of these kinds of drugs can inhibit this action. Cocaine and imipramine do not appear to release catecholamines into the synapses but do block reuptake. Cocaine and amphetamines have the same kinds of behavioral effects but imipramine does not. Imipramine in normals tends to be sedating. There is a latency period of several days before its antidepressant effects are observed. These are some of the grey areas in psychopharmacology.

One can argue that the several actions of tricyclic antidepressants like imipramine, complicate the picture and that perhaps they have some phenothiazine-like action. The M.A.O. inhibitors, which can also elevate catecholamines by inhibiting their enzymatic destruction, can in some cases give you an amphetamine-like psychosis. Almost all the amphetamine psychoses have paranoid features. And so you argue that these drugs release greater amounts of catecholamines and produce a schizophrenia-like psychosis, but why is it always paranoid? Does that mean paranoid schizophrenia is different from other forms of schizophrenia? One of the possible explanations that could account for the paranoid features of amphetamine psychoses is the fact that these drugs have multiple effects. At the simplest level, there are two major catecholamines in the brain, dopamine and norepinephrine. Let us say that amphetamines cause schizophrenia-like symptoms through the release of dopamine. Perhaps they do something else. They stimulate norepinephrine partially and there is evidence that the alerting action of amphetamines is related to norepinephrine. You might speculate that the difference between schizophrenia and paranoid schizophrenia is that the paranoid schizophrenic is a very alert schizophrenic. Amphetamines may, therefore, have a tendency to cause a schizophrenia-like syndrome through dopamine effects, while norepinephrine transforms this into a paranoid type of syndrome.

DR. LIPTON: Dr. Snyder pointed out the resemblance between the amphetamine psychosis and paranoid schizophrenia. There is general consensus that those two are difficult to distinguish. On the other hand, the hebephrenic schizophrenic and the simple schizophrenic are easy to distinguish from the amphetamine psychosis. Does this imply a different mechanism in these illnesses?

DR. GJESSING: Yes, it is difficult in chronic, paranoid, amphetamine psychosis, but it is also difficult in other kinds of intoxication.

DR. BLISS: I was wondering how you make the distinction?

DR. GJESSING: Well this is a clinical observation in acute amphetamine psychosis by Lars Gunne, Professor of Psychiatry at Uppsala. They had a wave of cases of acute amphetamine psychosis in Uppsala, and he had all of them coming to his clinic. I do not know all the details of his method of distinguishing these cases from "schizophrenia."

DR. BIGELOW: A common symptom in amphetamine toxicity is formication, a "crawling" sense under the skin. This is not usual in schizophrenia.

DR. ALTSCHULE: After big doses, for some reason, blood pressure and pulse rate may not be elevated. In fact they did not even show pupillary dilatation but they did have a peculiar sensation they all reported.

DR. BIGELOW: Formication?

DR. SNYDER: I hope Dr. Gunne evaluated his subjects blindly, that is without knowledge whether they had acute paranoid schizophrenia and/or amphetamine psychosis?

DR. GJESSING: No. He was not in a blind session. On the other hand, formication is well known in cocaine, paranoid psychosis.

DR. SNYDER: There are other concerns about the amphetamine psychosis as a model for schizophrenia. Observers differ in their interpretations -- some say that people with amphetamine psychosis do not have the classic schizophrenic thought disorder or disturbance of affect. Others disagree. One way of looking at it is that the paranoid process conceals the thought and affect disorders. A reasonably well organized paranoid individual does not appear as chaotic in his thinking process and a disturbance of affect may not appear as obvious as it would in a non-paranoid schizophrenic.

Methylphenidate (Ritalin) elicits a dramatic and florid exacerbation of schizophrenic symptoms in doses of only 29 mg. (95). The effect is not to superimpose an organic psychosis but to intensify the patient's own schizophrenic symptoms. If he is catatonic he becomes more catatonic, if hebephrenic, more hebephrenic.

Methylphenidate (Ritalin)

When we were talking about two catecholamines, dopamine and norepinephrine, we asked which of them might be more relevant. In this case, there is relevant evidence. The study I refer to employed both amphetamines and Ritalin. Ritalin is a more potent schizophrenogenic drug than amphetamine. But as a stimulant, amphetamine is more potent than Ritalin, and this is really interesting. It is an actual reversal. The potency in the worsening schizophrenic symptoms in comparing these two drugs is the opposite of the relative potency as stimulants.

To make a long story short, it looks like Ritalin acts preferentially on dopamine neurons as contrasted to amphetamines which act preferentially on norepinephrine. This suggests that the worsening of the schizophrenic symptoms by amphetamine like drugs may be mediated by dopamine as opposed to norepinephrine.

DR. GJESSING: With regard to the dopamine hypothesis, I would think of amphetamine first. As already pointed out, acute amphetamine psychosis is very different from schizophrenia. People who have been working with both acute amphetamine and schizophrenic patients can easily distinguish between them. So I am not so sure that data from acute amphetamine studies will support the dopamine hypothesis. Moreover although dopa is said to aggravate schizophrenic symptoms, I have not seen any worsening of the symptoms with the high dosage of dopa, as much as 13 grams of dopa. A third point is that monoamine oxidase inhibitor does not always aggravate schizophrenic symptoms which one would expect on the basis of the dopamine hypothesis.

The Relationship of Molecular Structure to Therapeutic Effectiveness

DR. SNYDER: I would like briefly to summarize the current theories that phenothiazines primarily block at dopaminergic receptors, and that the blockade correlates with their clinical effectiveness. For purposes of this discussion I do not propose to review all the available evidence.

The original suggestion that phenothiazines block at dopamine receptors stems from the early work of Arvid Carlsson (26) who studied the metabolism of dopamine. Careful scrutiny of the various metabolites of dopamine suggested that phenothiazines block dopamine receptor interactions in a dose dependent manner that correlated with their clinical potency. Such suggestions were strengthened by more direct studies of phenothiazine effects on catecholamine turnover.

More direct evidence was obtained by intracellular recordings from identified dopaminergic neurons, following iontophoretic application of dopamine (2). In dopaminergic neurons, dopamine is consistently inhibitory, and its effects are blocked by intravenous administration of phenothiazine in a dose dependent fashion. To date, the reversal of inhibition has not been done by iontophoretic application of phenothiazines.

I wish to stress that it is especially important to do the neurophysiology on cells that receive dopaminergic input, as that implies presence of dopamine receptors. Other, non-dopaminergic cells also respond to dopamine in a random fashion, much like the non-specific excitatory response of a variety of neurons to iontophoretic applications of aspartic and glutamic acids.

DR. KETY: What do you mean by phenothiazine reversal?

DR. SNYDER: Let me extend the dopamine model further, which will answer your question. The neurophysiology experiments are suggestive, but the technical difficulties preclude the testing of a variety of phenothiazine derivatives, which is necessary for meaningful structure-activity relationships. Paul Greengard at Yale observed that dopamine stimulates adenylate cyclase, and that the dopamine sensitive adenylate cyclase is enriched in areas rich in dopaminergic terminals and in dopamine (103). Thus the dopamine receptor is related to the adenylate cyclase, though it is not necessarily implied that the cyclase and receptor are identical moieties within the cell membrane. One might envisage that, in a fluid membrane, receptor and enzyme have certain mobilities and that formation of a complex between the dopamine and its receptor is permissive to interaction with the enzyme and stimulation by dopamine of intracellular cAMP accumulation.

Thus it is postulated that dopamine stimulates a specific adenylate cyclase, and that phenothiazine clinical efficacy is directly related to the degree of inhibition of the cyclase.

In the in vitro studies, phenothiazines have been shown to be potent blockers of the dopamine sensitive adenylate cyclase. The potency of inhibition is expressed as an inhibition constant K_I, which is the concentration necessary to occupy half the receptors.

Fluphenazine (Prolixin or Permitil) has a K_I of $8 \times 10^{-9}M$, and is the most potent phenothiazine marketed in the U.S. By and large Paul Greengard's data on correlations between clinical effectiveness and inhibition of the dopamine adenylate cyclase and other pharmacological tests are reasonable.

Fluphenazine

There are other drugs whose clinical effectiveness does not

correlate with enzyme inhibition studies. The most serious exceptions are the butyrophenomes, which in vivo are effective blockers of dopamine receptors but are unimpressive in in vitro test systems. Clinically, Haldol (Haloperidol) is more potent than Fluphenazine. However, it is a weak inhibitor of the cyclase. Direct labelling of the dopamine receptor with ^{3}H-dopamine and ^{3}H-haloperidol in our laboratory has clarified this dilemma. Dopamine and haloperidol respectively label agonist and antagonist states of the receptor. Affinity of both butyrophenomes and phenothiazines for haloperidol binding sites correlates very well with clinical potency (36).

Recently, Andy Feinberg and I have attempted to determine the possible fit of a variety of psychoactive drugs to the dopaminergic receptor, and correlate such data to clinical effectiveness (48). The antischizophrenic activity of phenothiazine drugs and their tendency to elicit extrapyramidal symptoms are thought to involve blockade of synaptic dopamine receptors in the brain. Space filling molecular models show how favorable Van der Waal's interactions between the side chain amino of phenothiazines and the 2-substituent on ring A can promote a conformation mimicking dopamine. These Van der Waal's attractive forces can explain: (1) the greater potency of drugs with trifluoromethyl rather than chlorine as a 2-substituent; (2) the enhanced activity of phenothiazines with piperazine instead of aklylamino side chains; (3) the increased potency associated with hydroxyethylpiperazines as contrasted to piperazine side chains; (4) the greater potency of cis rather than trans thioxanthenes; and (5) the crucial location of the ring A substituent at carbon no. 2. Potential energy calculations support the observations with molecular models and suggest an active conformation for the phenothiazines.

Differential rates of penetration or metabolic degradation of various drugs do not satisfactorily explain their differences in clinical effectiveness.

Considerations of drug conformational structure in comparison to dopamine, and therefore interactions with dopamine receptors provide a reasonable basis for structure activity relationships of psychoactive drugs. Therefore drugs that most clearly mimic the dopamine or phenylethylamine conformation are expected to be the most potent.

Metabolism of the butyrophenomes is not well worked out. However they are lipid soluble, so differential penetration into the brain does not account for the discrepancy.

DR. KETY: Nevertheless findings in vitro are not equivalent to those in vivo.

DR. SNYDER: I agree that the discrepancy exists only for the in vitro model.

DR. HABER: With reference to Dr. Kety's comments, the demethylated metabolites of chlorpromazine act more quickly than the parent compound, but the effects are shorter. The suggestion could be made that the demethylated derivatives are absorbed more rapidly from the G.I. tract, and significant tissue levels are more readily attained.

DR. SNYDER: I am not familiar with those studies.

DR. HABER: These are some old experiments I did with Harold Himwich on the intestinal absorption and tissue accumulation of chlorpromazine, imipramine and their mono- and di-demethylated derivatives. Regretfully, we never published those studies.

DR. LIPTON: In dealing with therapeutic effectiveness of the butyrophenomes and of the phenothiazines, is one more effective than the other in the treatment of various types of schizophrenia?

DR. SNYDER: Our qualitative impression is that there is little difference, though the data are rather general.

DR. LIPTON: We still have the dilemma of the long delay in the onset of therapeutic effects for these drugs. Slow rates of absorption may provide a partial explanation but it seems unlikely that many weeks would be required to achieve adequate tissue levels.

DR. SNYDER: I believe that there is little delay with the phenothiazines but most certainly there is for the tricyclic antidepressants.

Sedative versus Antipsychotic Effects of Phenothiazines

DR. LIPTON: No, I disagree. One must distinguish here between the sedative and antipsychotic effects of the phenothiazines. With regard to the latter, the phenothiazines act every bit as slowly as the antidepressants. An antipsychotic response usually takes more than a week to appear and an adequate therapeutic response about a month.

DR. BIGELOW: In fact, the onset of phenothiazine effectiveness as an antipsychotic may be even slower than tricyclics. Moreover, you often see prolonged antipsychotic effects for weeks after withdrawal.

DR. LIPTON: I agree with Dr. Bigelow. There is always a slow onset of clinical effectiveness of antipsychotic drugs but this is highly variable. There is even greater variability in the rate at which patients get ill after drugs are discontinued. So we have this problem of tremendous variability, both in rate of onset and in duration of antipsychotic effects of these drugs.

DR. BIGELOW: Everyone has observed tremendous variability in the recurrence of psychosis following phenothiazine withdrawal.

DR. CANCRO: It is my impression that sedative effects predominate during the first week or two of therapy with phenothiazines, and the antipsychotic effects are more apparent only after three or four weeks.

DR. HABER: Why is Prolixin such a long acting phenothiazine preparation?

DR. SNYDER: Long acting Prolixin is specifically formulated to ensure lipid solubility, and slow and sustained delivery. Its effectiveness is due to the formulation, and not differences in the metabolism of the drug per se.

DR. LIPTON: I agree.

DR. KETY: Overall the in vitro model may be far removed from the in vivo situation. You stated that in the amphetamine model you get good correlations between anti-amphetamine effects and therapeutic effects. Might it not be simpler to assume that the butyrophenomes and phenothiazines are metabolized differently, yielding pharmacologically active metabolic intermediates?

DR. HABER: Is it possible that receptor blockage by haloperidol feeds back on to dopamine synthesis, by analogy to tyrosine hydroxylase?

DR. SNYDER: Yes, that is true.

DR. HABER: If that is the case perhaps haloperidol, which has a low affinity in the in vitro system, is effective in vivo because of lesser feedback stimulation on tyrosine hydroxylase.

DR. SNYDER: Regretfully, the opposite is true.

DR. HABER: Then we still have a paradoxical situation to explain.

DR. COHEN: Could you tell us some details of the receptor binding assay.

DR. SNYDER: All the receptor assays involve the binding of radioactive ligands to particular brain preparations. The brains are homogenized in a hypertonic medium to rupture identifiable organelles and synaptic terminals. The preparation is then treated with a detergent to further disrupt the membranes. Following incubation with the radioactive ligand, you can quantitate the binding by ultrafiltration or centrifugation.

DR. WOLF: Could someone bring us up to date on the serotonin story?

Serotonin

DR. BIGELOW: There was much laughter in some circles when Dr. Woolley stated, "The trouble with schizophrenia is that there is not enough serotonin" and while not expecting it, we may have evidence that he was right all the time.

DR. GJESSING: Here I would like to mention a strange case studied in Lyon dealing with Tryptophan and 5-HO-Tryptophan. Dr. Catherine Fischer-Perrudon (51) investigated a case over eleven months. Her patient was in his twenties and was observed to go without sleep at all for three or four months. He went to bed every night but remained awake all the time. He had continuous EEG recordings day and night. He hallucinated in the evening between 8 and 10 p.m., but it only lasted twenty minutes to one hour and it was a most unpleasant experience for him. He was otherwise psychologically normal. He was diagnosed as Morvan's syndrome or Choree fibrillaire de Morvan. It is a very rare neurological condition with paresthesia, diarrhea, and muscular fibrillations. After having observed him for a long time many many kinds of hypnotics were tried without effect. Then he was given tryptophan in large amounts, 2-12 g. a day, also without benefit. The hallucinations even increased. When we finally gave him 5-HO-tryptophan, he reacted immediately with normal sleep. They kept him on 2-10 g. of 5-HO-tryptophan for three months. The increase in sleep time was directly related to the 5-HO-tryptophan doses. The hallucinations disappeared. But he became more and more depressed. When the 5-HO-tryptophan was discontinued, his insomnia returned. After two weeks on placebo they wanted to try him with 5-HO-tryptophan again, but it did not work this time and he died shortly thereafter.

DR. WYATT: L-5 HTP probably increased REM sleep a little in normal subjects. In contrast, L-tryptophan has a very profound soporific agent. In schizophrenic patients, large dosages of 5-HTP actually produces insomnia.

Enzyme Measurement

We studied many of the enzymes involved in serotonin metabolism. Prior to and during this study we also measured several other enzymes. Wise, Baden and Stein studied dopamine beta hydroxylase (DBH) the enzyme which converts dopamine to norepinephrine (190). They studied a series of 18 brains from chronic schizophrenics who had died in a state hospital and matched them with a group of other kinds of deaths that occurred in New York City. They found that there was a very marked deficit in this enzyme's activity. They attempted to control for most of the obvious artifacts and did it very well. We were collecting autopsied brains for another purpose and, when we heard of their findings, we tried to replicate them. In Dr. Jack Barchas' laboratory at Stanford we measured DBH in the brains of 9 controls and 9 schizophrenics (196). The difference between the schizophrenics and controls was very small. We can be criticized because our numbers are not as large as those of Wise and Stein, but their difference quantitatively was much greater than ours. There are some possible reasons for this. I want to talk about two correlations that could, perhaps, explain the small nonsignificant differences that we found. The first is time. In the best of hospitals it may take up to a few hours after death for the body to reach the morgue. Certainly in hospitals like Saint Elizabeth's and other state hospitals, which are short-staffed, it can take many hours. Also, in large psychiatric hospitals, autopsies are not usually performed until many hours after the body reaches the morgue. We made correlations between the enzyme activity and the time it took after the body reached the morgue to the time of autopsy and it was fairly high in an inverse fashion.

In the Wise and Stein experiments the material was collected in two different locations. Their controls were collected in New York City, where there is probably a very active morgue with quick autopsies, whereas in the state hospital where the schizophrenics came from there may be some delay. The other significant correlation we found came from comparing the dose of chlorpromazine or its equivalent that the patient was taking when he was last observed by his physician and the enzyme activity. Again, there was a significant negative correlation between the DBH activity and the dose of drug. Wise and Stein did try to control for this by giving rats chlorpromazine chronically for about three months and did not find any decrease in enzyme activity.

DR. COHEN: You would still have endogenous protease activity which might increase due to the time factor.

DR. WYATT: Right. In Barchas' laboratory we are taking autopsied normal brains and we are letting them sit at room temperature for various time periods. This, of course, approximates what happens to our patients' bodies. We then keep them at 4 degrees for periods of time. The variance in enzyme activity with these procedures appears to be large, but we will talk more about this when we have concluded this experiment.

DR. GJESSING: In animals, is there a difference between biopsy and autopsy?

DR. WYATT: No, not in rats that either we or Wise and Stein have studied, but with respect to the human material, there may be a discrepancy between what we are told when we get the brains and what, in fact, actually happens. We are told that the bodies reach the morgue in two or three hours. However, there may actually be a delay of eight to ten hours.

Suppose a patient dies during the night shift in the hospital where there is one nurse on the ward or one nursing attendant. The patient dies at 11:00 p.m. There is no way of getting that body to the morgue until the next morning.

DR. KETY: The patient may have been pronounced dead five hours late.

DR. WYATT: Right.

DR. KETY: How about the nonschizophrenic?

DR. WYATT: The time of death in the nonschizophrenics is probably a little clearer because autopsies on most of these are fairly urgent -- car accident cases, guns, etc. -- but in the case of our schizophrenics, we cannot be certain about the hour of the death. I am not sure Wise and Stein can either. It is still possible, however, the deficits are real, in which case they are very exciting. We will not know until we get a good idea of how long it takes for enzymes to decay in the human brain.

DR. KETY: You did not do other enzymes?

DR. WYATT: Yes, we have some other enzymes.

DR. KETY: In the same brain?

DR. WYATT: Yes.

DR. KETY: Were they also decreased?

DR. WYATT: Yes, for example, phenylethanolamine-N-methyltranferase was also decreased. The one in which we were especially interested is the dopamine specific adenyl cyclase. We measured it in the head of a caudate and it was normal (25). The response to stimulation by dopamine was also normal. So here we have an enzyme that seems normal in the two sets of brain. Unfortunately we know that in rat brains there is a 66% drop off of activity in three hours and then it plateaus; there is no further drop. We are probably starting off with low levels in the schizophrenic brain, but if there is a difference, you might see it with stimulation. This did not occur when the enzyme was stimulated by dopamine.

Another enzyme of interest is monoamine oxidase (MAO) (10, 11, 27, 132, 134, 135, 136, 195, 196, 197). We were initially interested in it because of a possible abnormality in indoleamine metabolism which we were thinking about. If there were a block in MAO presumably there would be a build-up of indoleamines as well. There would also be an excess of dopamine, which would fit in with the general hypothesis that Dr. Snyder presented.

MAO Activity

The assay I am going to discuss was done in Dr. Dennis Murphy's laboratory at the National Institute of Mental Health. We used the conversion rate of radioactive tryptamine in an homogenate of platelets to indoleacetic-aldehyde. We have not found that moderate dosages of phenothiazines alter the mean enzyme activity.

DR. HABER: Were these last cases matched for age?

DR. WYATT: Yes, with regard to the brain assays. The mean ages were extremely close for the schizophrenics and controls. Although this was certainly desirable, the closeness in age was more by chance than design. We now have over 308 normals for MAO activity (Fig. 3). They have a reasonably normal frequency distribution. The unipolar depressed patients collected by Dr. Murphy are no different from the normals. For acute schizophrenics, and I will not at this point try to define what I mean by acute schizophrenics, again, there is NO difference. When Dr. Murphy examined a group of bipolar depressed individuals, there was a shift toward the left. The mean was 3.6 versus 5.0 for the normals. Then we come to chronic schizophrenic patients, hospitalized for several years, who are phenothiazine refractory in the sense that they cannot leave the hospital even with drugs. As a group they are also very low.

If we simply look at the number of patients who are below 2 units of MAO, we find about 10% of normals (30 out of 308), 2% of unipolar patients; none of the acute schizophrenics; 20% of the bipolars; and 40% of the chronic schizophrenics (Table XIV). With simple statistics (chi square), you can see that there is a big difference between the chronic schizophrenics and the normals. The difference does not come out to be statistically significant between the bipolars and the normals, or between the bipolar and the schizophrenics. It may be that when we have some slightly larger numbers of subjects a difference will appear, but right now we are not able to say whether the bipolars fit into the normal group or whether they fit more into that of the schizophrenics.

Dr. Murphy found that the correlations of platelet MAO in normal monozygotic twins was .88 whereas for dizygotic twins the correlation was .45 (133), suggesting the enzyme was under genetic control. Wondering how to quickly resolve the issue of potential artifact in the chronic schizophrenics, and knowing that the enzyme

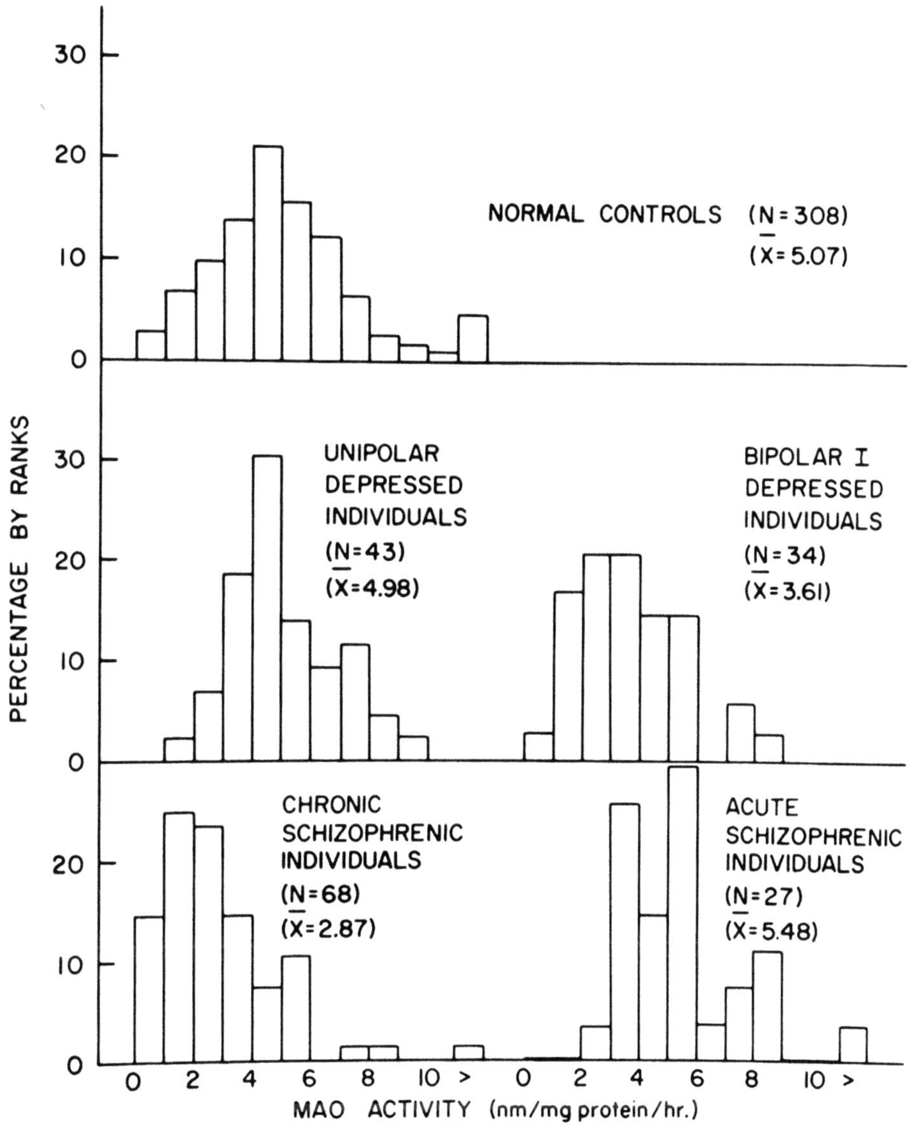

FIG. 3. Distribution frequency of platelet MAO activity in normal, chronic schizophrenic, acute schizophrenic, unipolar depressed, and bipolar I depressed individuals.

TABLE XIV

	MAO BELOW 2 UNITS
NORMALS	30/308
UNIPOLAR	1/43
ACUTE SCHIZOPHRENICS	0/27
BIPOLAR I	7/34
CHRONIC SCHIZOPHRENICS	28/68

activity is pretty well controlled by genetic factors in normal people, we decided to study the sets of monozygotic twins discordant for schizophrenia accumulated by Dr. William Pollin at the National Institute of Mental Health. They had scattered over the United States and Mexico, but we were able to contact 13 sets living in 23 different cities or communities. We went around and took blood samples from the schizophrenics and from anyone standing around who looked approximately normal. We coded them, sent them back to Washington, and ran them in double blind fashion.

DR. KETY: What did you do at the time that you took the blood from the patients at home? Did you centrifuge it? Did you coagulate it or what?

DR. WYATT: We asked the subject if he would come to a nearby facility, hospital or laboratory. If he refused, we took the blood in his home, put it on ice, ran to the nearest centrifuge we could find in whatever community we were in, spun it down and froze it under dry ice, and at the same time we grabbed the normals.

DR. KETY: You spun it down and separated it?

DR. WYATT: Yes.

DR. KETY: And then froze it?

DR. WYATT: Yes. I think you can see that there is a general reduction in enzyme activity in the schizophrenic twin, as well as in his nonschizophrenic co-twin, with a correlation of .67 between the twins. You will also see that the twins seem to have a bimodal distribution; four sets of twins are very low and the others are in the general normal range (Fig. 4). We did not recognize this at the time we did the study, but it turned out that all four of the low twin sets had a chronic schizophrenic index.

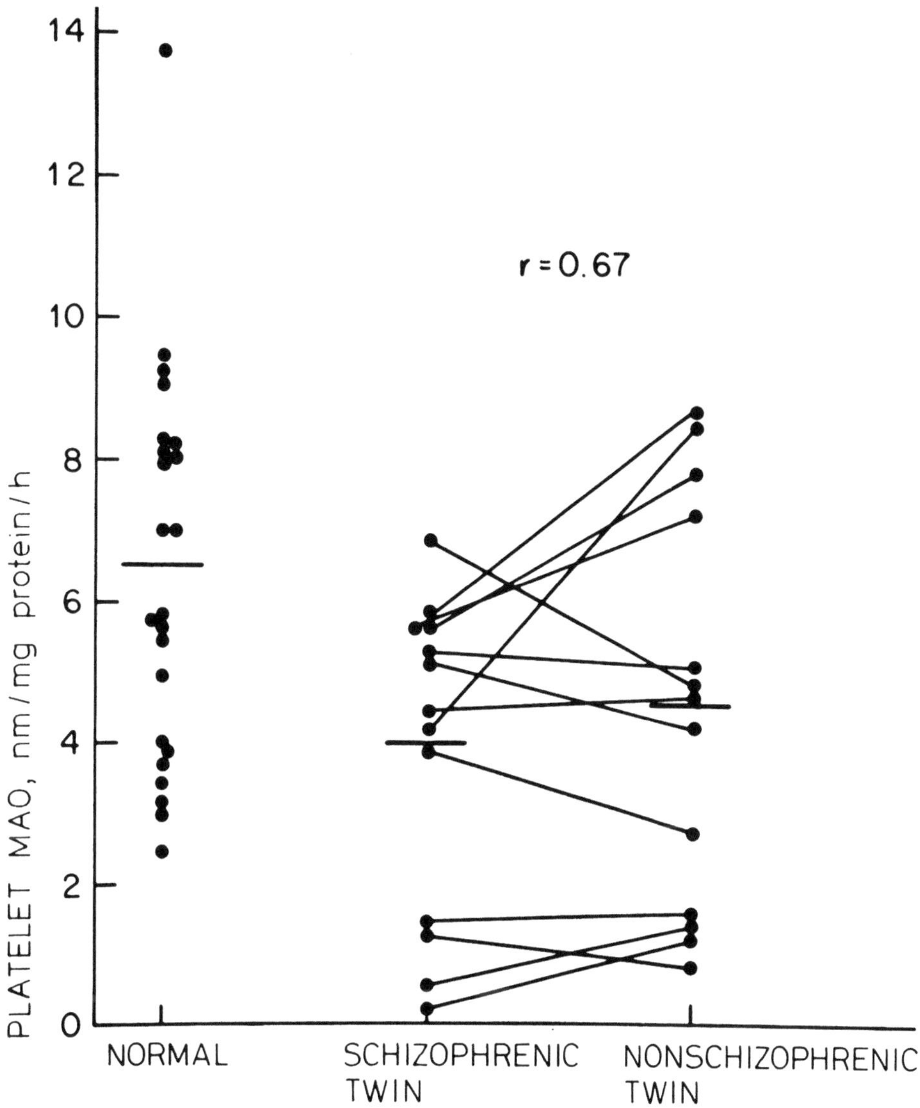

FIG. 4. MAO activity in 23 normal and 13 pairs of monozygotic twins discordant for schizophrenia. Correlation is between discordant twins (p 0.01).

DR. KETY: And what were the rest?

DR. WYATT: A mixture. They are about half acute and half chronic.

DR. KETY: But there are some good chronics among the others?

DR. WYATT: Yes. Not every chronic schizophrenic has a low MAO.

DR. KETY: But when you find a low MAO, you are very likely to find a chronic schizophrenic.

DR. WYATT: Not necessarily. There are about 10% of normals with a low MAO. One day Dr. Monte Buchsbaum at N.I.M.H. is going to tell us who those normals are. He has been studying several hundred of them with sophisticated psychological and physiological tests.

DR. KETY: You mean there are people among the normal population who have low MAO?

DR. WYATT: Yes.

DR. KETY: And you find then that they are not normal?

DR. WYATT: Some of them are not normal and some of them may be perfectly normal.

DR. KETY: You mean that a low MAO is a reasonably good predictor of psychopathology even in a normal population?

DR. WYATT: We do not have that broken down, but we looked at some of the normals and expect to report in greater detail on this in the next year. But some of the "normal" 10% whose MAO activity is under 2 units are not really normal.

DR. KETY: But you have reason to think that the low MAO group have a higher incidence of psychopathology.

DR. WYATT: We think that is probably the case. The point of the twin study is that there are nonschizophrenic, identical twins who have never been exposed to drugs, hospitalization, or a peculiar diet, but still have the low MAO. Low platelet MAO is therefore not simply an artifact of being on phenothiazines or being stressed.

To see whether the low MAO might have a functional significance, we (59) have also measured platelet serotonin in the same group of chronic schizophrenic patients. We selected serotonin because it was easy to measure. Serotonin concentrations were slightly elevated in the patients. In the patients given phenothiazine the difference disappeared, presumably because the phenothiazines prevent serotonin from being taken up into the platelets.

Platelet MAO versus Brain MAO in Schizophrenia

The platelets, of course, are in the blood and nobody has ever thought the platelets were responsible for schizophrenia. What about the brain? What is brain MAO like in schizophrenics? Fortunately, MAO is stable. So long as the tissue is not broken up, there is little decay with time (160). There is no difference between the enzyme activity in the brains of our 9 controls and our 9 schizophrenics, using the tryptamine assay used with the platelets. It turns out that there may be two kinds of enzymes. For convenience they are called type A and type B. The type A enzyme is a good substrate for beta-phenylethylamine. The enzymes can be selectively inhibited; type A by deprenyl, type B by pargyline. We also examined MAO type A and type B in these brains. Again, there was no difference between schizophrenics and controls (161).

Is Schizophrenia a Disease of the Brain or a Systemic Metabolic Disease that alters Brain Function

How then does one rationalize this kind of dilemma. There is an abnormality in the platelet in a disorder which we consider to be a disease of the brain. Among the possibilities that should be considered is an analogy with phenylketonuria (PKU). PKU, of course, produces very marked changes in the brain, but is not a disease of the brain at all; it is a disease of the liver. In schizophrenia we thought that we ought to be studying the brain and that is what we studied, but maybe we ought to be studying the liver, the lungs, the kidneys, and the muscle as well as other places where MAO is located.

Let us assume for a minute that it is established that there is a deficit of monoamine oxidase in the body of schizophrenics. What would be the implications of that? MAO is an enzyme that is used to metabolize a large number of substrates, not only the monoamines but also the diamines. Jumping ahead, how can you test which one of these might be the one involved if, in fact, MAO is low and causative? There are a couple of ways of doing this.

The Effects of Dimethyl Tryptamine

We have set up a series of test criteria which I will go through with one substance and tell you about some preliminary experiments we are doing with another substance which I think are fairly exciting. Tryptamine is a substrate for MAO. With a block in MAO, tryptamine could be converted to dimethyltryptamine (DMT), a known hallucinogen. The criteria were initially set up by Hollister (85); we have modified them a bit (194).

1. The first criterion is that the agent be able to mimic some of the aspects of schizophrenia. We have now given dimethyltryptamine (DMT) to 20 normal volunteers. They had all taken hallucinogenic drugs many times previously and planned to take them again. The dosage was 0.7 mg. per kilogram. DMT produces a very short acting, subjective high. The subjects withdrew into themselves and did not want to communicate with us. They did not have frank visual hallucinations, but they had some visual distortions. With DMT there was some of the general flavor of what might take place in a schizophrenic -- the withdrawal, the interest in self and not in the outside world, the unwillingness to communicate or the inability, whichever. There was, however, no formal thought disorder and there were no auditory hallucinations. DMT was short-acting; the high, occurring very quickly, was gone within an hour. It can be called the businessman's trip, because you can take it over your lunch period and then go back to work!

2. The second criterion is that "the agent must be found in man." I think now that there is very good evidence, primarily produced by Dr. Lewis Mandel and the group at Merck (183). Using gas liquid chromatography, DMT can be found in extraordinarily low concentrations in human tissue.

3. The agent's precursor must be found in man. I think it is now established through studies like those of Savaadra and Axelrod (158) that tryptamine is present in human tissues.

4. The fourth criterion is that "the agent must be synthesized in man." In the early 1960's the first reports came out from Dr. Axelrod's laboratory (8) showing that the lung of a rabbit could convert tryptamine to a dimethyltryptamine. It is now clear that the human lung can also convert tryptamine to dimethyltryptamine (117) in vitro, but there is no evidence whether this occurs in vivo.

5. The fifth criterion is "the agent must be differentially synthesized, or metabolized in schizophrenics." This criterion has clearly not been met to date. No good study has demonstrated unequivocally that there are increased levels of DMT in any constituent of the body in schizophrenics versus controls.

6. Tolerance should not develop. There are a number of studies indicating that tolerance does not develop to DMT. One study was done by Dr. Gillin (64). He gave DMT and LSD to cats with implanted electrodes. With LSD, there were some high voltage slow waves, which dissipated by the second or third day. With DMT there was no change in the EEG or in the behavior of the animal after 15 consecutive days of administration. Dr. Jonathan Kaplan (101) did a similar experiment with four of the human subjects we had studied previously. He gave DMT twice a day for five days and again found that no tolerance

developed. This is quite different from the human experience with LSD where there is complete tolerance by the second or third day of continued administration.

7. One criterion that has not been tested is whether the neuroleptic drugs block the effects of DMT, although we do know they block some of the hallucinogens.

MAO as a marker in schizophrenia

DR. BLISS: You have an observation that platelet MAO is low in some chronic schizophrenics and that platelet MAO tends to be stable. Furthermore you have a suspicion that this is in part genetically determined. If it all worked out you would have a marker of something which is stable and low in some schizophrenics. Now, I want to ask about the stability. If it turned out that 50% of all people destined to become schizophrenics had a low MAO even if that were not etiological, it would be a lovely thing to have. Is there any reason to believe that it does not have this degree of stability?

DR. WYATT: We know it does fluctuate, for example, with the menstrual cycle.

DR. BLISS: How much?

DR. WYATT: About 23%, peak to trough. That is about all we do know that will really influence this other than specific MAO inhibitors.

DR. BLISS: What about drugs?

DR. WYATT: This is a problem which we are just getting into, in normals the fluctuation is fairly small, and I think we can say it is about 17% over a period of several years where we changed our assay numerous times.

DR. BLISS: Diurinal variations?

DR. WYATT: No, there is no diurinal variation.

DR. BLISS: What about people who are sick with other illnesses, such as nephritis, etc?

DR. WYATT: It does not seem to vary. We have looked at a group of people with various endocrine diseases and they do not seem to show increased variability in platelet MAO.

DR. ALTSCHULE: How about alcoholics?

DR. WYATT: We tested a lot of them and when we go back to test them they do not always come out the same way they did the first time.

DR. LIPTON: When the level of MAO rapidly changes by 100% there is no clinical change, is there?

DR. SNYDER: I always wondered about reports that platelet levels of serotonin are abnormally low in mongolism. Many, many laboratories found that platelet serotonin was low, I think, in childhood schizophrenia or autism.

DR. WYATT: After listening to Dr. Snyder present his construction of the dopamine hypothesis and amphetamine for several years, something dawned on me. This is the structure of amphetamine (Fig.5). No one has proposed that amphetamine is an endogenous substance, and therefore it probably is not responsible for schizophrenia. But there is a substance very much like amphetamine, phenylethylamine (PEA) which is clearly endogenously formed. It is a direct metabolite of phenylalanine and is elevated in patients with PKU. The relatives of patients with PKU are said, in some of the literature, to have a very high incidence of psychosis. Well, of course, phenylethylamine also turns out to be the best substrate we know for the B type MAO which is the type of enzyme that is low in the platelets. One might expect to find an excess of PEA or one of its many metabolites. PEA can be hydroxylated in several places. It goes to phenylethyanolamine, it can be hydroxylated on the ring, and probably that ring hydroxyl can be methylated.

Phenylethylamine

DR. GJESSING: I have given MAO inhibitor to a 30 year old patient with clinical PKU. His I.Q. was about 50 or so. And he was having some behavior trouble on the ward, not very much, but after an MAO inhibitor, he improved. There were no signs of psychosis at all.

DR. HABER: Just a couple of comments about a relative of PEA, that is paramethoxyphenylethylamine (PMPEA), which might possibly be found in the urine, and has some very interesting behavioral effects. In vivo, it produces a catatonic-like rigidity which lasts about 10 or 15 minutes. You can repeat the experiment again and again.

DR. WOLF: This is in rats?

DR. HABER: Yes. Physiologically PMPEA has all kinds of interactions with monoaminergic pathways involved in the modulation of spinal reflexes (6, 188). In vitro, PMPEA seems to preferentially block the transport of 5-HT and, to a lesser extent, that of NE and DA (5). Like phenylethylamine, this is an indirectly acting sympathomimetic amine, but differs from PEA in that PMPEA seems to release

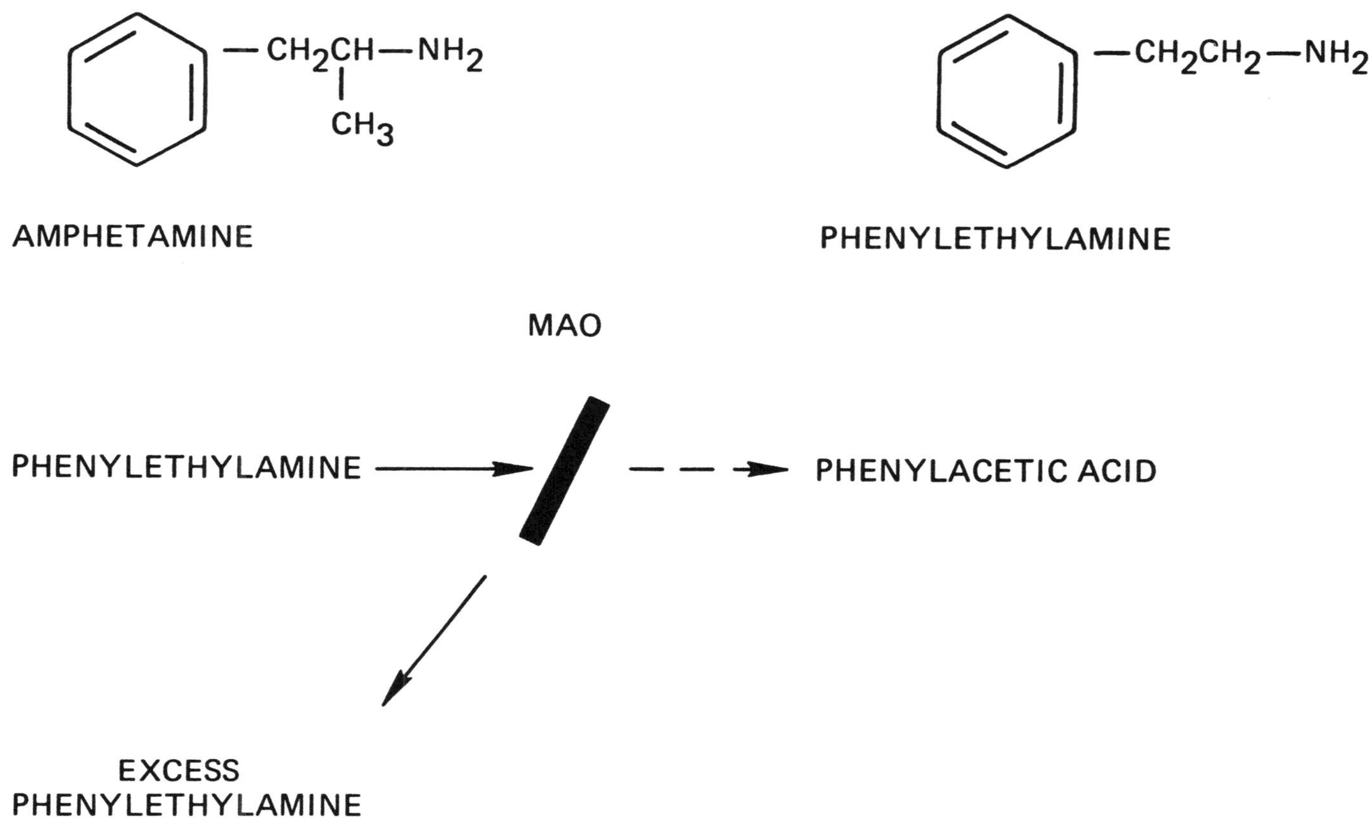

FIG. 5. The structures of amphetamine and the endogenous related compound, phenylethylamine. Also shown is how a hypothetical block in phenylethylamine metabolism by decreased monoamine oxidase (MAO) could increase phenylethylamine.

preferentially serotonin and norepinephrine but not dopamine (7). Dr. Ashkenazi in Israel has recently observed that Haloperidol, at 0.5 mgs. per kg., blocked the PMPEA effects on head rigidity but not on the leg rigidity. It is clear that the monomethylated PE (PMPEA) is much more effective than either the parent amine or the 2,3 and 3,4 dimethoxy derivatives. It would seem that in vivo compounds like PMPEA might well be formed from PE, which does occur in the central nervous system.

DR. KETY: There are a large number of metabolites of these drugs. All it takes is one of those metabolites to be different. How about paramethoxy PEA, which you have not looked for?

DR. WYATT: There are a number of metabolites of PEA. The other one we looked at is phenylethanolamine, but now we have to look at the hydroxylated ring compounds which we have not started to look at yet.

DR. KETY: You are going to look for the methoxylated derivatives?

DR. WYATT: We will look for both the hydroxylated and methoxylated.

DR. LIPTON: In rats we tried to optimize conditions for elevating DMT by administering large doses of tryptophan, pyridoxin, an MAO inhibitor and methionine. We still got no rise in DMT, but the animals became very sick, their temperature went to about 108 degrees and they died in about three or four hours. Yet we could still not detect DMT.

DR. WYATT: I think DMT is present in some human tissues in very low concentrations, tenths of a nanogram per ml. of blood. It might be in our food. We do not know if it is actually made in our bodies.

DR. HABER: In those studies where those patients were getting that tryptophan and 5 HTP, did you get a change in DMT?

DR. WYATT: We measured one patient given L-tryptophan and DMT was not altered.

DR. BLISS: If I recall your description of the effects of the drugs, it did not sound very much like schizophrenia, is that correct?

DR. WYATT: It certainly does not sound like chronic schizophrenia.

DR. KETY: It does not sound like LSD, does it? It does not sound like what Szara described?

Effects of Hallucinogens

DR. WYATT: No. It is not exactly like Steve Szara described it. We do not find visual hallucinations. We do not generally find auditory hallucinations. We found subjective changes, withdrawal and a couple of frank paranoid reactions where people became very, very suspicious and did not want to have any thing to do with us. Fortunately, the paranoia lasts only for the length of drug administration.

DR. WOLF: And you get electroencephalographic changes, is that what you said?

DR. WYATT: Those were in cats. We have not looked in man.

DR. SNYDER: I collected a lot of anecdotal information on drug culture in San Francisco when the drug epidemic was at its height. I have taken DMT and I have taken LSD. DMT is the same as LSD except that it lasts only 45 minutes to an hour. My impression is that for the psychedelic drugs this tolerance is a very characteristic effect and that there is cross-tolerance, even between mescaline and LSD. So the use of these drugs by people has been useful in ascertaining the possibility of their acting via the same mechanism. Tolerance is classically determined by dose and duration. And since this tolerance and cross-tolerance is characteristic of all psychedelic drugs, I would strongly suspect that the failure to develop tolerance for the DMT is just because it is very short-acting. This is something, I suppose, that could be studied in animals if you just loaded them with DMT, in order to keep the level up for a long time.

DR. WYATT: Isbell did a study with LSD (153) where he gave people LSD until they become tolerant, then gave them DMT and there was little cross-tolerance to DMT. We gave DMT to cats every two hours for 48 hours and no tolerance developed, which is more along the model which you would like to see. Dr. Domino (107) has done this a little differently. He used a FR4 schedule giving DMT to rats every two hours for two weeks; half the animals became tolerant and the others did not. I think we can say, in general, that there are a certain group of people or animals who do not become tolerant to this drug.

DR. LIPTON: We have seen subjects who are cross-tolerant to mescaline and psilocyamine but not to DMT.

CHAPTER VI - Clinical Investigations

Participants: Drs. Altschule, Bigelow, Gjessing, Lipton, Snyder Wolf, Wyatt.

DR. BIGELOW: There is a growing consensus that the term schizophrenia stands for a syndrome most probably of diverse etiologies. If the syndrome is defined rigorously and specifically enough, there is a high probability that most psychiatrists anywhere in the world would agree on the diagnosis. Although we may thereby be able to narrow the concept and facilitate diagnostic agreement, we are not thereby assured that the patients so labeled will be homogeneous from an etiologic standpoint.

Granting, therefore, that we have no a priori way of selecting an etiologically homogeneous group of schizophrenics for study, what procedures should we follow in setting up clinical experiments? In the Laboratory of Clinical Psychopharmacology, part of the NIMH Intramural Research Program housed at Saint Elizabeth's Hospital, we have opted to study selected chronic schizophrenic patients individually and on a relatively long-term basis. The patients are young, voluntary and obviously refractory to conventional treatments or they would not be referred to the research ward. They are symptomatically active. We try to exclude from our sample patients who appear to be suffering mainly from the social and psychological consequences of long term hospitalization and social isolation, although, of course, we have no rigorous criteria on which to make this exclusion. It follows from their agreement to come to the ward and try new medications that they have retained some optimism, some internal sense that things could be better, however magical or unrealistic such an attitude might seem to the more informed or sophisticated observer. Our experience in obtaining experimental consent from the patients is that almost none regard themselves as "ill," but most do acknowledge that they could feel better and agree, on that basis, to participate in the clinical trials. The matter of "informed consent" is a complex one which I will not attempt to deal with here.

In our Laboratory we have focused much of our energy on trials of new biological treatments. For measurement of clinical status we have used a nurses' rating scale, currently in its "second generation" (Fig. 6). The items were developed in cooperation with a nursing staff of varied educational and professional background and represents to some extent a compromise between what we wished to measure and what emerged as possible to rate with clarity, confidence and agreement. It can be seen that the scale covers a fairly wide range of expressed psychopathology. Unlike the NOSIE (87) there are no items representing positive achievement. Staff

NIMH INPATIENT BEHAVIORAL RATING SCALE

PATIENT:	I.D. NUMBER	DATE: MONTH DAY YEAR	DAY OF YEAR:
	0 1 2 3 4 5 6 7 8 9	M 0 1	0 1 2 3
	0 1 2 3 4 5 6 7 8 9	M 0 1 2 3 4 5 6 7 8 9	0 1 2 3 4 5 6 7 8 9
	0 1 2 3 4 5 6 7 8 9	D 0 1 2 3	0 1 2 3 4 5 6 7 8 9
RATER:	I.D. NUMBER	D 0 1 2 3 4 5 6 7 8 9	HOSPITAL STUDY CODE:
	0 1 2 3 4 5 6 7 8 9	Y 7 8 9	0 1 2 3 4 5 6 7 8 9
	0 1 2 3 4 5 6 7 8 9	Y 0 1 2 3 4 5 6 7 8 9	0 1 2 3 4 5 6 7 8 9
	0 1 2 3 4 5 6 7 8 9	SHIFT: DAY EVE	0 1 2 3 4 5 6 7 8 9

Rate TYPICAL Behavior: 0=NOT PRESENT; 1,2=MILD; 3,4,5=MODERATE; 6,7=SEVERE.

1 MENTAL ILLNESS-GLOBAL Functional impairment; unable to cope w problems	0 1 2 3 4 5 6 7	2 EUPHORIA, HIGH MOOD Overly cheerful;Grandiose; Dramatic;Show-off;Elated	0 1 2 3 4 5 6 7
3 DISRUPTIVE IRRITABILITY Complaining; Criticizing; Demanding; Blames others	0 1 2 3 4 5 6 7	4 HYPERACTIVITY, AGITATION Increased motor activity; Restless; Overly reactive	0 1 2 3 4 5 6 7
5 ANGER-STATEMENTS Angry, abusive comments, Threats; Profanity; etc.	0 1 2 3 4 5 6 7	6 ANGER-BEHAVIORAL Appears hostile; Sullen, Threatening; Assaultive	0 1 2 3 4 5 6 7
7 ANXIETY-STATEMENTS Anxious, worrying, tense statements	0 1 2 3 4 5 6 7	8 ANXIETY-BEHAVIORAL Appears anxious, tense; Seems worried, frightened	0 1 2 3 4 5 6 7
9 DEPRESSION-STATEMENTS Sad, unhappy, worthless, or guilty comments; etc.	0 1 2 3 4 5 6 7	10 DEPRESSION-BEHAVIORAL Appears sad, unhappy, or "blue"; Seems depressed	0 1 2 3 4 5 6 7
11 DEPENDENCY Constantly seeks advice; Clinging helplessness	0 1 2 3 4 5 6 7	12 DECREASED ACTIVITY Slowed down or inactive; Seems tired, lethargic	0 1 2 3 4 5 6 7
13 CONFUSION, DISORIENTATION Difficulty concentrating, remembering, recognizing	0 1 2 3 4 5 6 7	14 REPETITIVE BEHAVIOR Same phrase, activity, or ritual done over and over	0 1 2 3 4 5 6 7
15 LOSS OF EMOTIONS Blunted feelings; Flat tone; Seems apathetic	0 1 2 3 4 5 6 7	16 DISORGANIZED SPEECH FORM Rambling, incoherent, or fragmented speech pattern	0 1 2 3 4 5 6 7
17 DELUSIONS-STATEMENTS Statements of illogical, "crazy" beliefs, thoughts	0 1 2 3 4 5 6 7	18 BIZARRE BEHAVIOR Unusual or "crazy" acts, gestures, outbursts, etc.	0 1 2 3 4 5 6 7
19 HALLUCINATIONS-STATEMENTS Talks about or to voices or other hallucinations	0 1 2 3 4 5 6 7	20 HALLUCINATIONS-BEHAVIORAL Gestures at, Preoccupied w apparent hallucinations	0 1 2 3 4 5 6 7
21 PARANOID DISTRUST Appears suspicious; Feels people are "against" him	0 1 2 3 4 5 6 7	22 SOCIAL ISOLATION Tends to be alone; Avoids participation; Seclusive	0 1 2 3 4 5 6 7
23 UNCOOPERATIVENESS Resents, resists, or refuses tasks, meds, etc.	0 1 2 3 4 5 6 7	24 SOCIAL INCOMPETENCE Lacks manners;Crude; Puts others off; Inconsiderate	0 1 2 3 4 5 6 7
25 SEXUAL PREOCCUPATION Appears aroused; Frequent overtures, talk, gestures	0 1 2 3 4 5 6 7	26 APPEARANCE Messy; Unkempt; Sloppy; Poor hygiene; etc.	0 1 2 3 4 5 6 7
27 SECLUSION ROOM Hours (0-7)	0 1 2 3 4 5 6 7	28	0 1 2 3 4 5 6 7

NOTES:

PLEASE USE A NO. 2 PENCIL. MAKE MARKS HEAVY.
ERASE COMPLETELY ANY MARK YOU WISH TO CHANGE.

FIG. 6

cannot "reward" any patient by assigning positive points, as it were. In this sense the scale is oriented only to record pathology. This may be a serious flaw in the scale, but we found during the development of the form that nursing staff had difficulty shifting their quantitative focus from rating pathology to recognizing assets. Thus errors were introduced.

The scale differs from the NOSIE further and more importantly in that many of the questions are more general and encompass a greater range of behaviors. We have found that nursing personnel are able to grasp the degree of abstraction inherent in some of the questions and that they find such questions more meaningful than the more specific ones in the NOSIE.

The scale is completed for each patient by two raters on each shift, day and evening. They trade information on what has been observed during that shift, but the selection of actual score -- ranging from zero not present, to 7, most severe -- is selected independently by each rater. Inter-rater reliabilities for these questions are good. After a period of familiarization and training we obtained the correlation coefficients shown in Table XV. That the concepts and observations of the staff are relatively stable in relation to the questions is indicated by the great degree of stability shown in the profile of clinically stable patients when rated on two occasions separated by two months (Fig. 7).

There are two particular facets of the scale that deserve special mention. Several of the areas addressed have two nearly identical questions. One item will ask for the rater's interpretation -- how the patient "seemed", and the other of the pair asks if there is specific mention of that symptom made by the patient. Thus, anxiety, anger, depression and hallucinations are all items asked both from the point of view of what the rater infers and from the observation of what the patient actually said. This has proven to be a distinction readily made by nursing staff. Most significantly there are often low correlations between the statement and behavior sections of these items.

None of the questions could be shown to correlate highly with any other question across even so small a sample as seven patients. Although there are many instances of high correlations, which is not surprising, there are also low correlations. This result indicates that there are at least no extraneous items on the scale, that is there are no items that could be shown to be redundant or tapping the same behavioral entity.

In developing the scale, we at first divided many questions into two parts. One score was assigned to reflect the extreme degree of a given symptom -- such as anger -- and the other to

TABLE XV

INTRACLASS CORRELATIONS:
ESTIMATES OF THE RELIABILITY OF THE MEAN
OF TWO MEASUREMENTS FOR RATING SCALE ITEMS

ITEMS	WARD 1		WARD 2	
	R_1	R_2	R_1	R_2
1. Mental Illness - Global	0.62	0.77	0.81	0.90
2. Euphoria	--	--	0.35	0.51
3. Disruptive Irritability	0.82	0.90	0.71	0.83
4. Hyperactivity	0.78	0.88	0.71	0.83
5. Anger - Statements	0.89	0.94	0.58	0.73
6. Anger - Behavioral	0.96	0.98	0.53	0.69
7. Anxiety - Statements	0.68	0.81	0.61	0.76
8. Anxiety - Behavioral	0.38	0.56	0.72	0.83
9. Depression - Statements	0.86	0.92	0.44	0.61
10. Depression - Behavioral	0.84	0.92	0.52	0.68
11. Dependency	0.91	0.95	0.65	0.79
12. Decreased activity	0.49	0.66	0.32	0.48
13. Confusion	0.58	0.73	0.56	0.72
14. Repetitive Behavior	0.76	0.87	0.49	0.66
15. Loss of Emotions	0.40	0.57	0.67	0.80
16. Disorganized Speech Form	0.68	0.81	0.56	0.72
17. Delusions - Statements	0.52	0.68	0.76	0.86
18. Bizarre Behavior	0.57	0.73	0.84	0.91
19. Hallucinations - Statements	0.53	0.69	0.80	0.89
20. Hallucinations - Behavioral	0.32	0.48	0.84	0.91
21. Paranoid Distrust	0.87	0.93	0.68	0.71
22. Social Isolation	0.84	0.91	0.61	0.76
23. Uncooperative	0.84	0.91	0.29	0.45
24. Social Incompetence	0.75	0.86	0.52	0.68
25. Sexual Preoccupation	0.90	0.95	0.56	0.72
26. Appearance	0.82	0.90	0.27	0.43
27. Seclusion Room	--	--	--	--

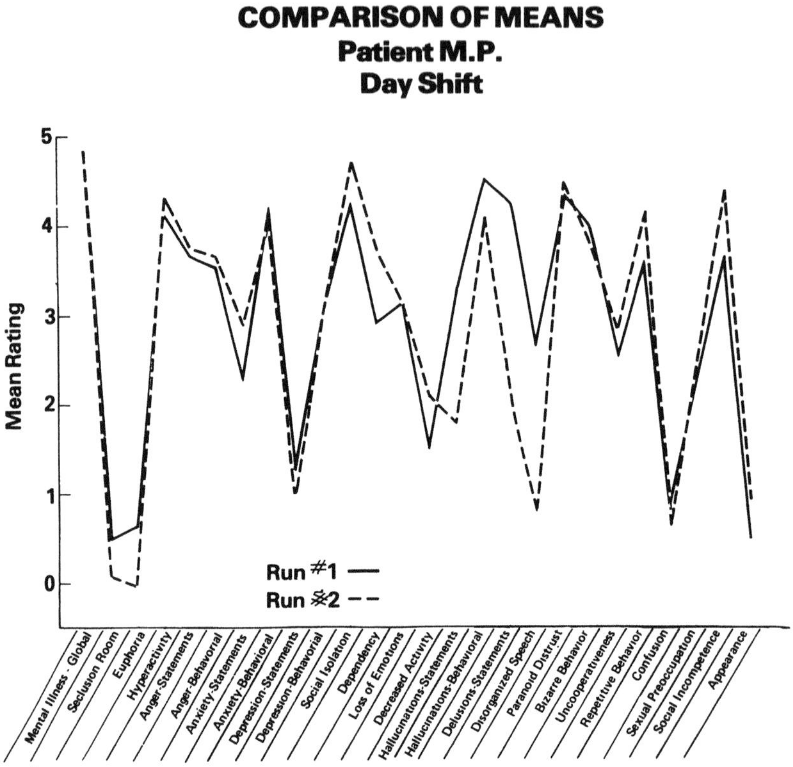

FIG. 7

reflect the judgment of the rater of the typical state for that shift. It was predicted that this scheme would permit raters to assign extreme scores for short-lived symptoms which appeared only briefly, and to assign a lower score to reflect the average state for that patient. This bifurcation was introduced partly at the request of the raters who had felt a need for an opportunity to make such a distinction. Despite the presence of a number of clinically volatile patients, this distinction did not prove to be a meaningful one. Correlations between typical and extreme scores were very high, indicating that we were obtaining no additional information from such a doubling of asked for responses. The idea was dropped. This experience duplicates that of the depression research units in Bethesda.

This scale obviously generates an enormous number of numbers. Each patient is rated four times a day. Four times 26 questions equals 104 responses a day. A three month protocol generates 90 times 104 responses or 9,360 data points. We are just beginning to develop the capability for utilizing this quantity of information. Our initial analyses have focused around the questions most closely allied to psychotic phenomena. To estimate the degree of psychosis present in a patient on a given shift the averaged response to six questions is summed and this sum is called the "psychosis rating" for that patient. The six questions are:

1. Loss of emotions;
2. Disorganized speech form;
3. Delusional statements;
4. Bizarre behavior;
5. Hallucinations - statements;
6. Hallucination - behavioral.

The degree of constancy or stability of any one of these items, or indeed the sum of the items, will show greater or lesser variation depending on the patient and depending on what is going on on the ward. In order to discover trends over longer periods of time -- particularly as relates to experimental manipulations -- we subject the data to a smoothing process. We compute the moving average. To do this for, say, a 10 day moving average, we find the average of day 1 - 10 and graph that point over day 5. The point to be plotted for day 6 is the average value for days 2 through 11; that for day 7 is the average rating from days 3 through 12 and so on. There is no particular magic compelling one to use a 10 day average; any other number may be used. The visual effect of taking moving averages is shown in Figs. 8, 9.

Figs. 8, 9 show the effects of L-5-hydroxytryptophan administration to a 26 year old man, ill for the past ten years and unresponsive to milieu and pharmacotherapy. He spent his days preoccupied with the dimension of things, his and other's feet, their height and weight. At worst, he was mute -- nearly catatonic. At his best, he was an extremely rigid person, holding his body in hyperextension and walking about in a highly stylized manner. His speech was confined to skeletal phrases delivered in a whisper almost reminiscent of the mock Oriental style of Charlie Chan. He drew pictures of guns endlessly, detailing their characteristics. No specific pattern of change in his psychosis score is evident from either the daily score or the moving averages (Fig. 8). Note also that the graphs from the day and evening shifts (Fig. 9) are nearly parallel but that the evening shift score is about 30% less than the day total.

You will note that there is a straight line drawn through the graphs. This line is the linear regression of all the points in the graph. Setting up this equation and line can be particularly

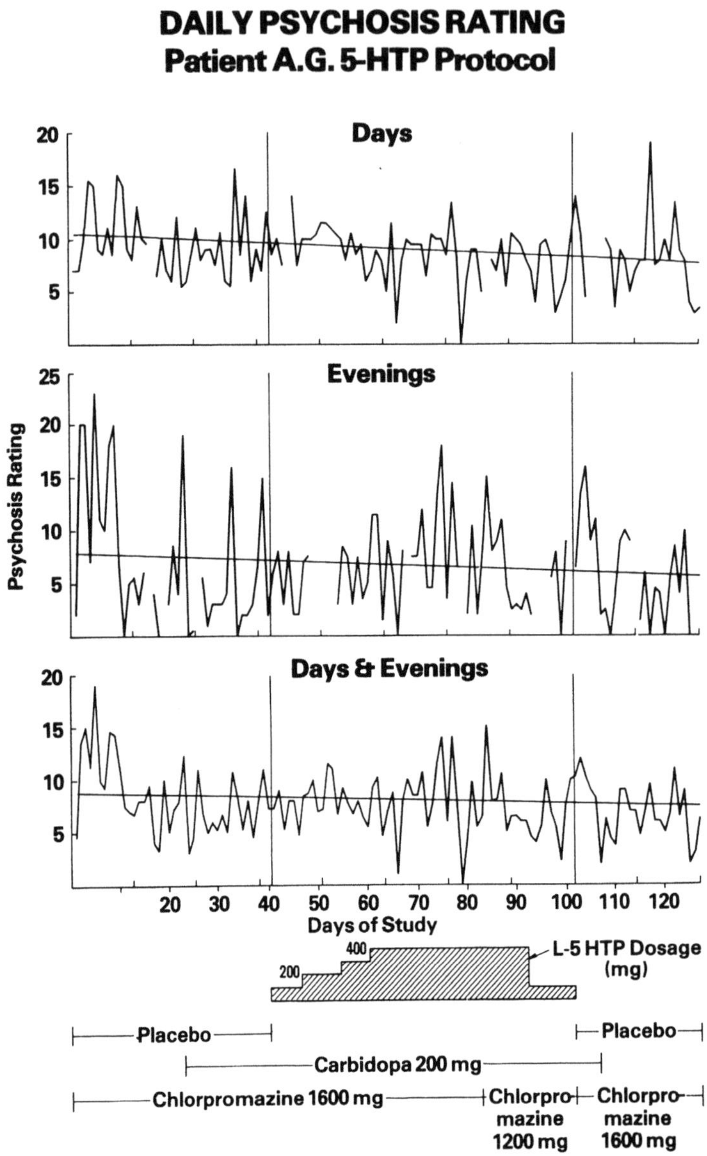
DAILY PSYCHOSIS RATING
Patient A.G. 5-HTP Protocol
Days
Evenings
Days & Evenings
Psychosis Rating
20
15
10
5
25
20
15
10
5
20
15
10
5
20 30 40 50 60 70 80 90 100 110 120
Days of Study
400
200
L-5 HTP Dosage (mg)
Placebo
Placebo
Carbidopa 200 mg
Chlorpromazine 1600 mg
Chlorpromazine 1200 mg
Chlorpromazine 1600 mg

FIG. 8

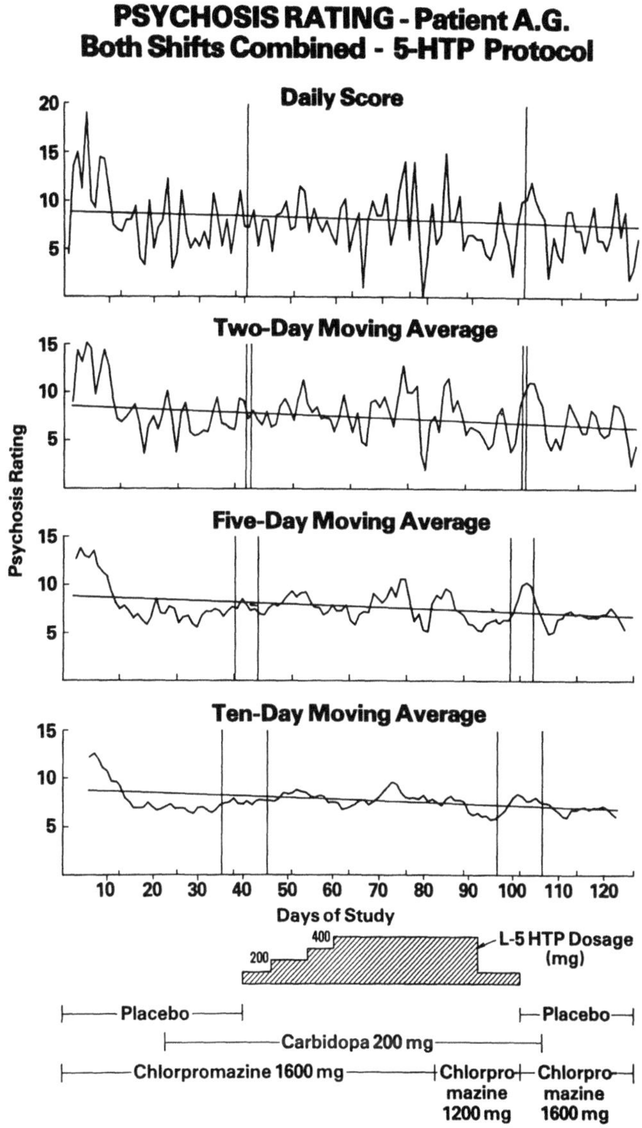
PSYCHOSIS RATING - Patient A.G.
Both Shifts Combined - 5-HTP Protocol
Daily Score
Two-Day Moving Average
Five-Day Moving Average
Ten-Day Moving Average
Psychosis Rating
Days of Study
L-5 HTP Dosage (mg)
400
200
Placebo
Placebo
Carbidopa 200 mg
Chlorpromazine 1600 mg
Chlorpromazine 1200 mg
Chlorpromazine 1600 mg

FIG. 9

useful in clinical studies when patients are undergoing long-term trends to improvement or deterioration. One can make statistical comparison of the effect of treatments by using as primary values the distance above or below the regression line. The weakness of such an approach is that there is, of course, no theoretical way of demonstrating that long-term trends are linear. In fact, they really cannot be linear or at some point the line would intercept zero in the case of improvement, or go on to a level higher than that mathematically possible in the case of worsening.

The next patient's graphs to be considered are those of a 19 year old woman, hospitalized for the past three years. She had prominent perceptual symptoms, stating frequently that she did or did not feel horses jumping from her eyes. In this instance the 5-hydroxytryptophan was associated with a marked exacerbation of her illness. The Psychosis score nearly doubled (Fig. 10). The graphs of some of the responses to individual questions are shown in Fig. 11. This figure shows that the L-5HTP was associated with an abrupt onset of confusion and a doubling of ratings for the Delusional Statements. The increase in rated Bizarre Behavior was less striking but clearly significant. Note that although the experimental medication was discontinued after 10 days the ratings did not return to baseline levels for almost a month. Results of this sort take a toll on patient, staff and researcher.

The graphs of the next patient are similarly derived from the L-5-hydroxtryptophan protocol. The patient is a 26 year old woman who had been hospitalized for most of the preceding 10 years. She was a diminutive, underweight person desperately involved in attempts to find someone to extricate her from a personal world filled with hallucinated horrors such as giant insects, and malevolent persons who tortured her, her family and acquaintances. Poisons were perceived everywhere and she pleaded with any passer-by to rectify the horrors she experienced as ongoing events. She had no insight into the unreality of these perceptions. Figs. 12 through 15 illustrate the daily 2, 5, and 10 day moving averages of her psychosis score. The results for each shift are presented separately. The daily score is fairly noisy but the significant trends are brought into sharp focus by the moving averages; the effect of the drug is shown clearly.

The day shift ratings show little change but the evening shift ratings indicate gradual improvement to a point followed by apparent worsening as the dose of 5-hydroxytryptophan was increased.

Figs. 16, 17, 18 show the 2 day moving average of some of the individual questions from the evening shift. These are included to illustrate both the clinical changes observed and to demonstrate the split in ratings of observed statements versus behavioral inference.

A more conventional approach to ratings using weekly NOSIES or

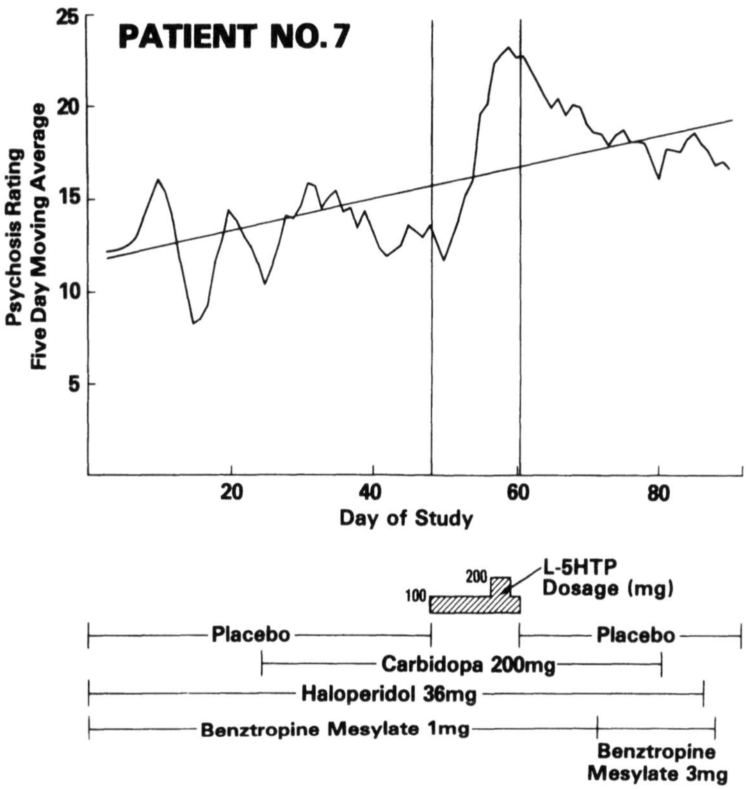
PATIENT NO. 7
25
20
15
10
5
Psychosis Rating
Five Day Moving Average
20
40
60
80
Day of Study
200
100
L-5HTP
Dosage (mg)
Placebo
Placebo
Carbidopa 200mg
Haloperidol 36mg
Benztropine Mesylate 1mg
Benztropine
Mesylate 3mg

FIG. 10

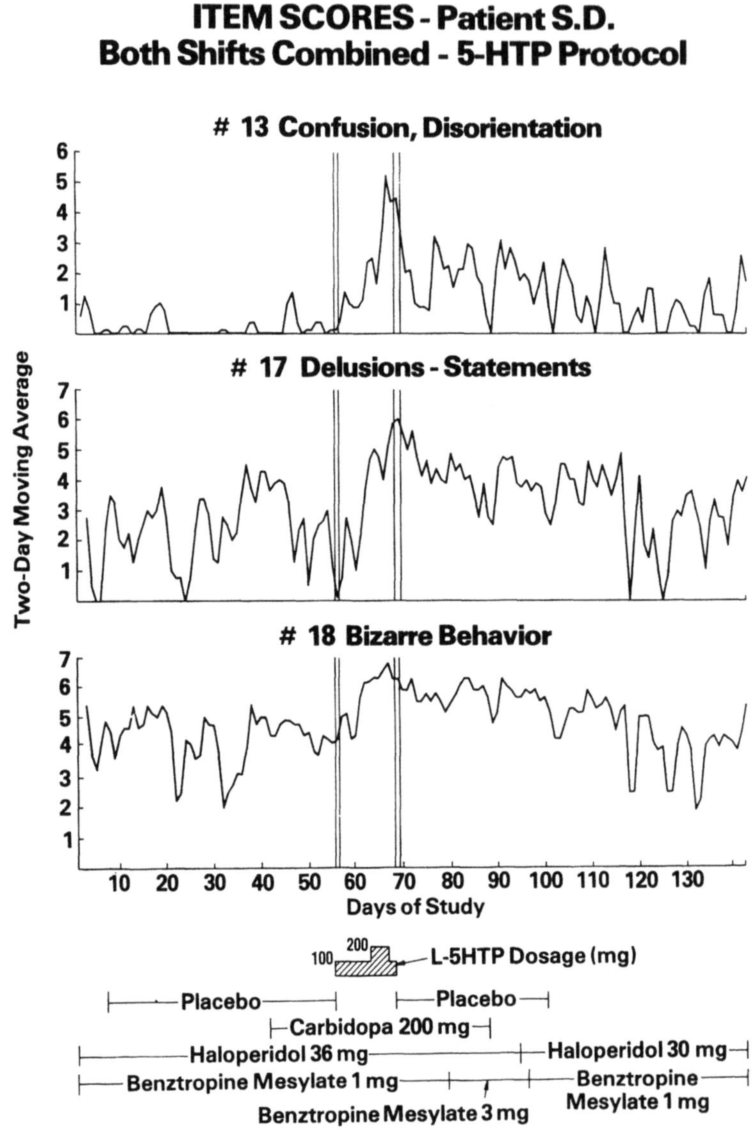
ITEM SCORES - Patient S.D.
Both Shifts Combined - 5-HTP Protocol
13 Confusion, Disorientation
17 Delusions - Statements
18 Bizarre Behavior
Two-Day Moving Average
Days of Study
L-5HTP Dosage (mg)
100
200
Placebo
Placebo
Carbidopa 200 mg
Haloperidol 36 mg
Haloperidol 30 mg
Benztropine Mesylate 1 mg
Benztropine Mesylate 1 mg
Benztropine Mesylate 3 mg

FIG. 11

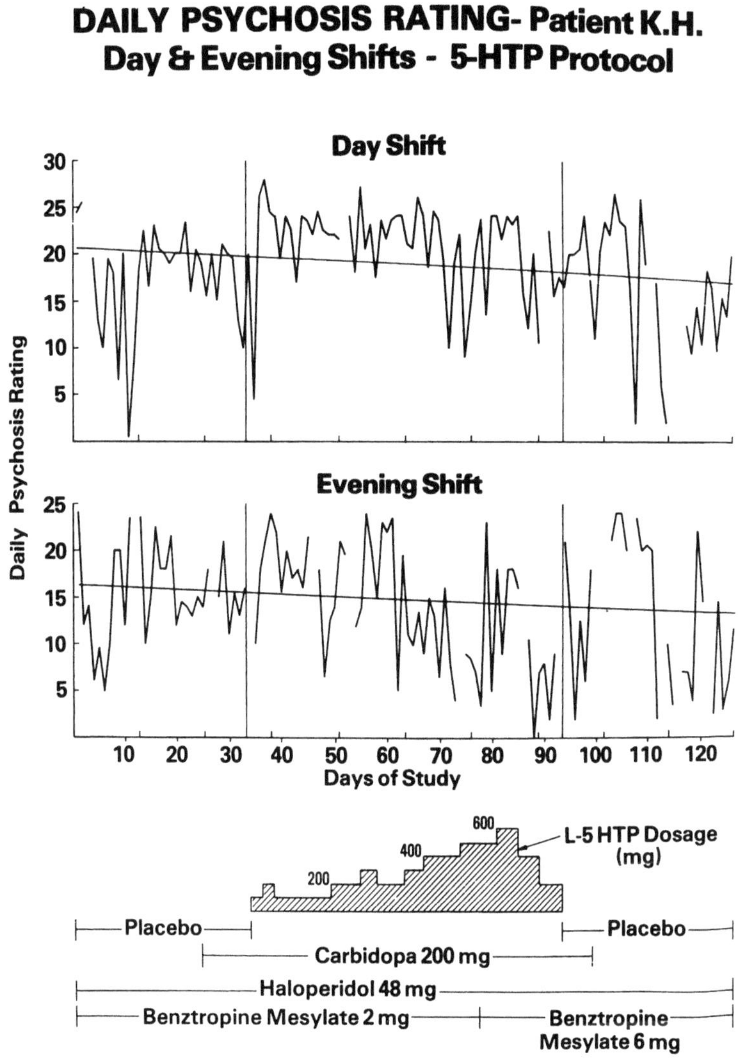
DAILY PSYCHOSIS RATING- Patient K.H.
Day & Evening Shifts - 5-HTP Protocol
Day Shift
Evening Shift
Daily Psychosis Rating
30
25
20
15
10
5
10 20 30 40 50 60 70 80 90 100 110 120
Days of Study
L-5 HTP Dosage (mg)
200
400
600
Placebo
Placebo
Carbidopa 200 mg
Haloperidol 48 mg
Benztropine Mesylate 2 mg
Benztropine Mesylate 6 mg

FIG. 12

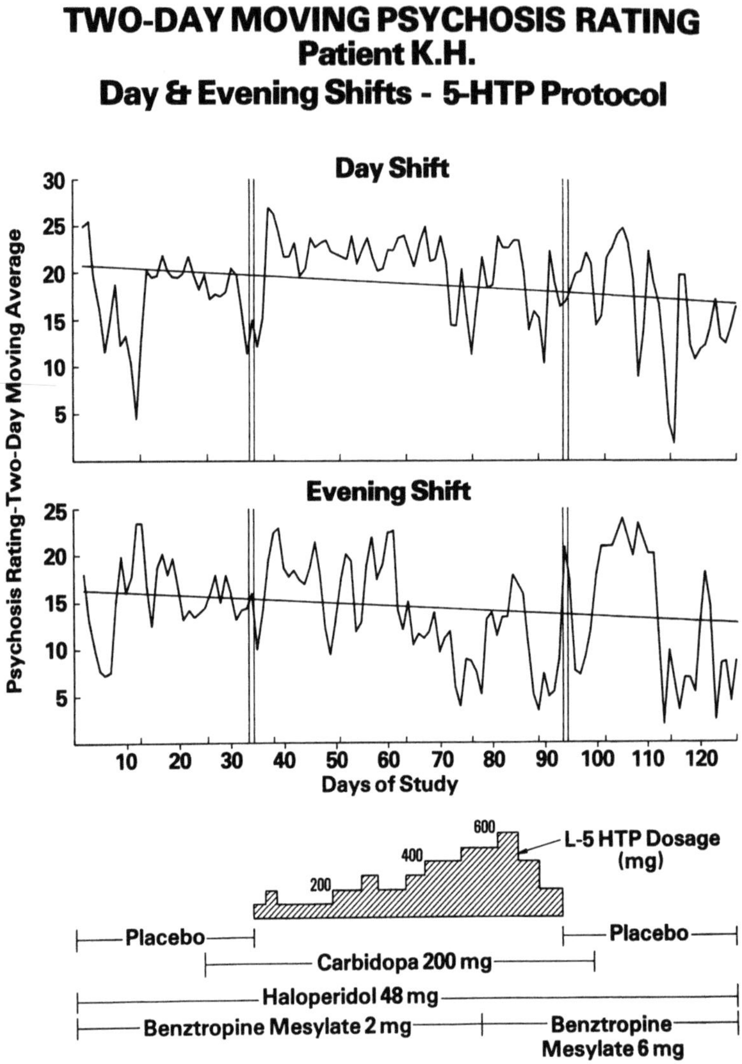
TWO-DAY MOVING PSYCHOSIS RATING
Patient K.H.
Day & Evening Shifts - 5-HTP Protocol
Day Shift
Evening Shift
Psychosis Rating-Two-Day Moving Average
30
25
20
15
10
5
10 20 30 40 50 60 70 80 90 100 110 120
Days of Study
200
400
600
L-5 HTP Dosage (mg)
Placebo
Placebo
Carbidopa 200 mg
Haloperidol 48 mg
Benztropine Mesylate 2 mg
Benztropine Mesylate 6 mg

FIG. 13

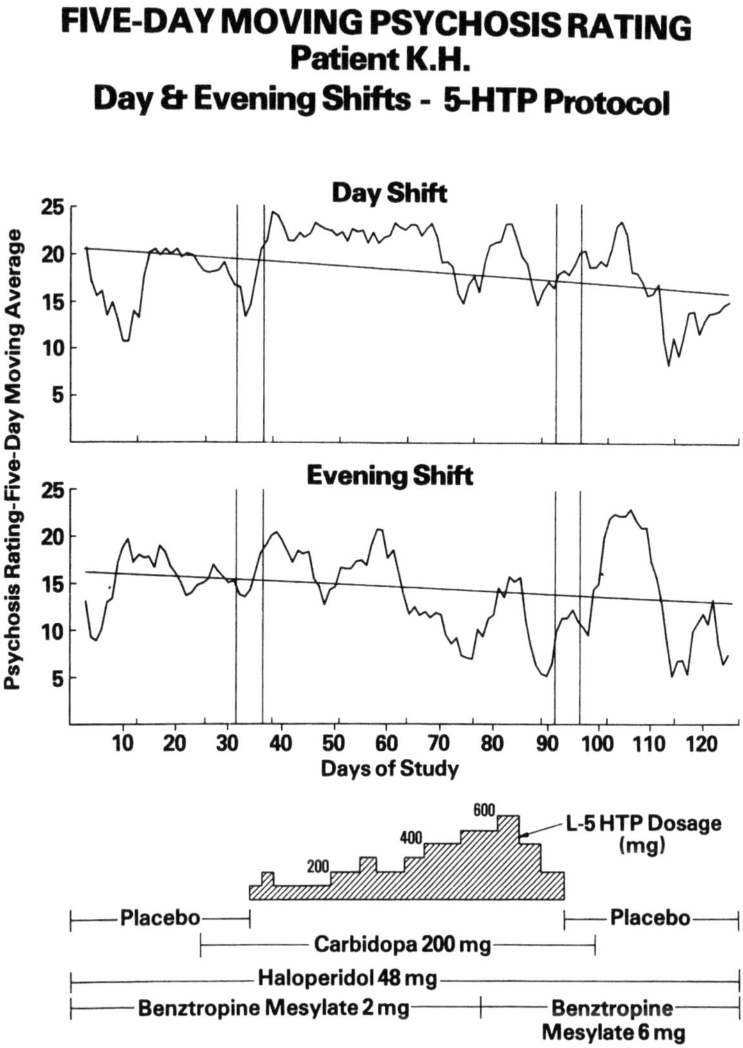
FIVE-DAY MOVING PSYCHOSIS RATING
Patient K.H.
Day & Evening Shifts - 5-HTP Protocol
Day Shift
Evening Shift
Psychosis Rating-Five-Day Moving Average
25
20
15
10
5
10 20 30 40 50 60 70 80 90 100 110 120
Days of Study
200
400
600
L-5 HTP Dosage (mg)
Placebo
Placebo
Carbidopa 200 mg
Haloperidol 48 mg
Benztropine Mesylate 2 mg
Benztropine Mesylate 6 mg

FIG. 14

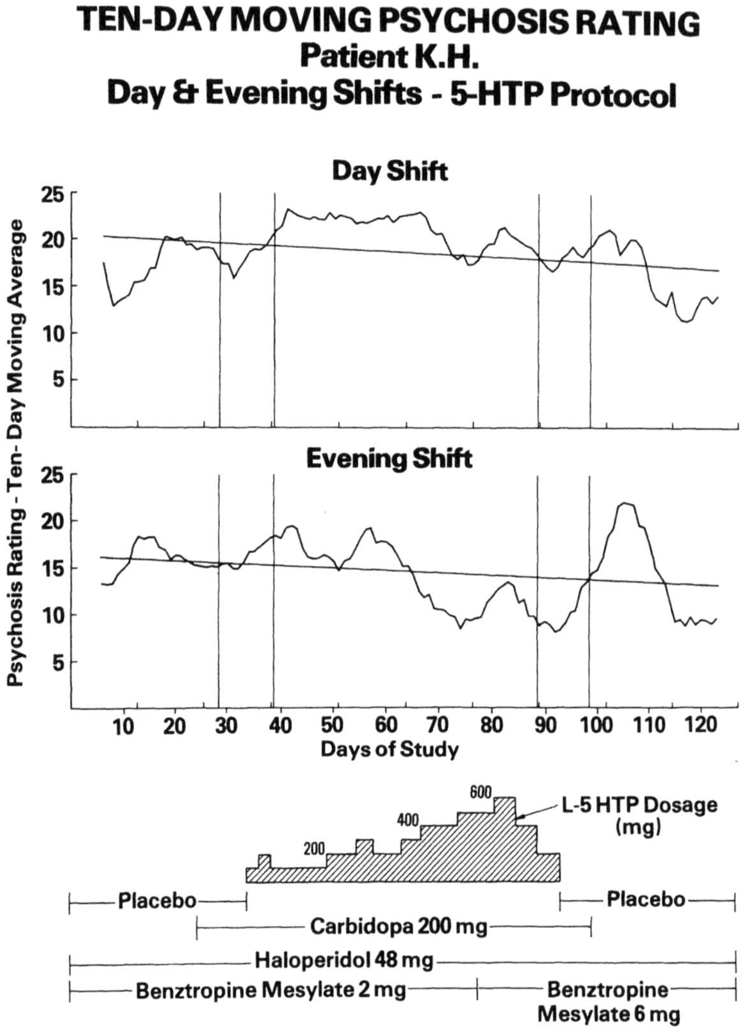
TEN-DAY MOVING PSYCHOSIS RATING
Patient K.H.
Day & Evening Shifts - 5-HTP Protocol
Day Shift
Evening Shift
Psychosis Rating - Ten- Day Moving Average
25
20
15
10
5
10 20 30 40 50 60 70 80 90 100 110 120
Days of Study
200
400
600
L-5 HTP Dosage (mg)
Placebo
Placebo
Carbidopa 200 mg
Haloperidol 48 mg
Benztropine Mesylate 2 mg
Benztropine Mesylate 6 mg

FIG. 15

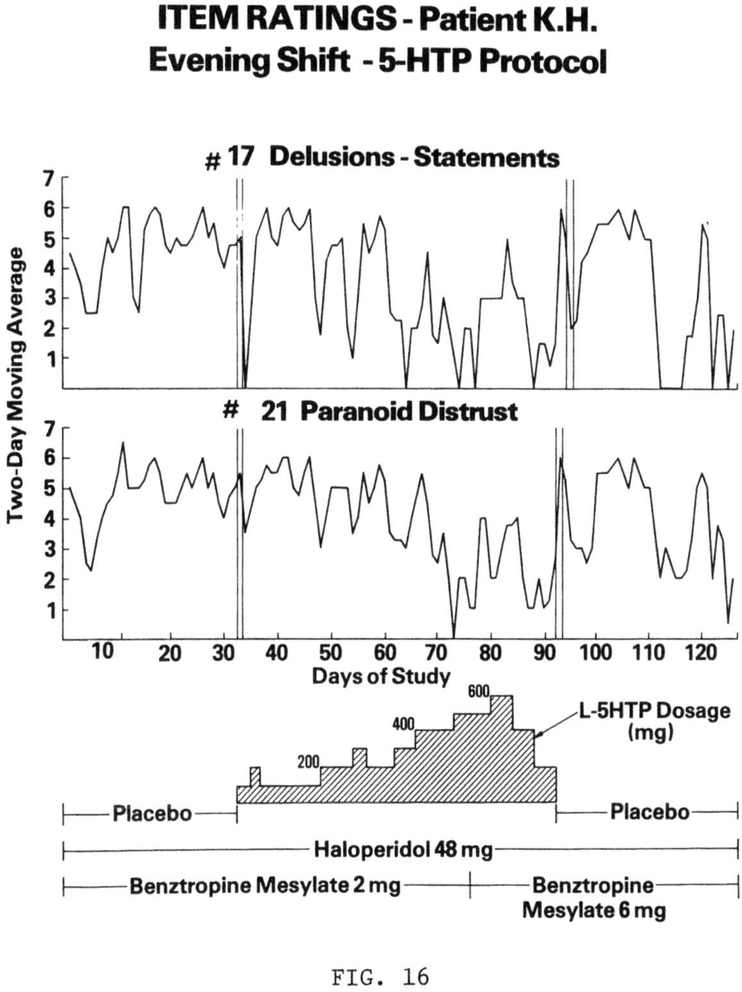
ITEM RATINGS - Patient K.H.
Evening Shift - 5-HTP Protocol
17 Delusions - Statements
21 Paranoid Distrust
Two-Day Moving Average
1
2
3
4
5
6
7
10
20
30
40
50
60
70
80
90
100
110
120
Days of Study
200
400
600
L-5HTP Dosage (mg)
Placebo
Placebo
Haloperidol 48 mg
Benztropine Mesylate 2 mg
Benztropine Mesylate 6 mg

FIG. 16

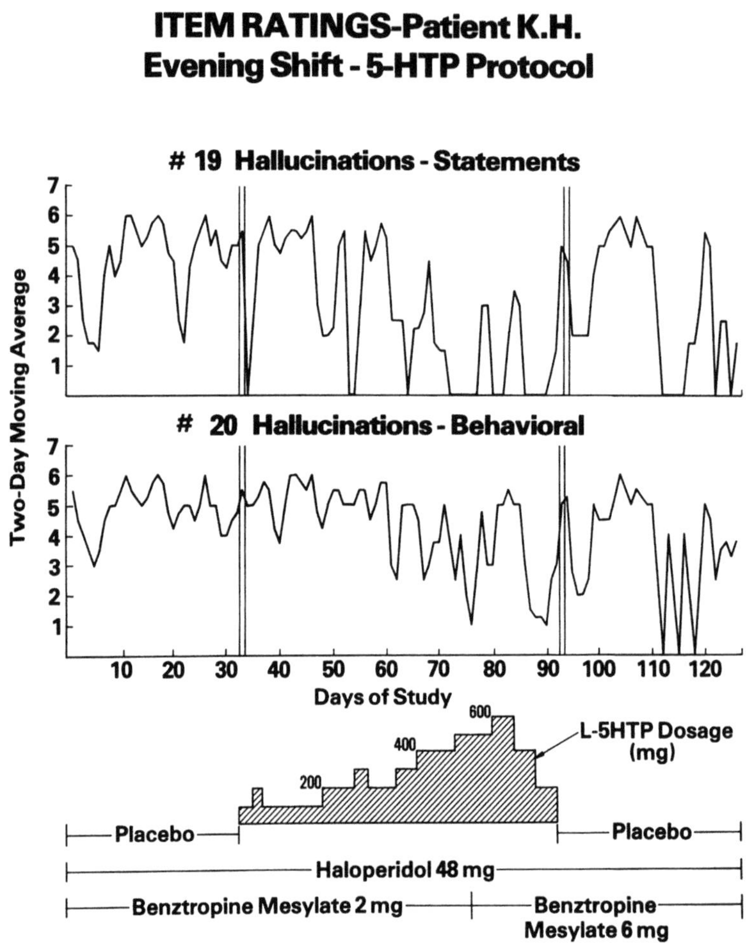
ITEM RATINGS-Patient K.H.
Evening Shift - 5-HTP Protocol
19 Hallucinations - Statements
20 Hallucinations - Behavioral
Two-Day Moving Average
1
2
3
4
5
6
7
10
20
30
40
50
60
70
80
90
100
110
120
Days of Study
200
400
600
L-5HTP Dosage (mg)
Placebo
Placebo
Haloperidol 48 mg
Benztropine Mesylate 2 mg
Benztropine Mesylate 6 mg

FIG. 17

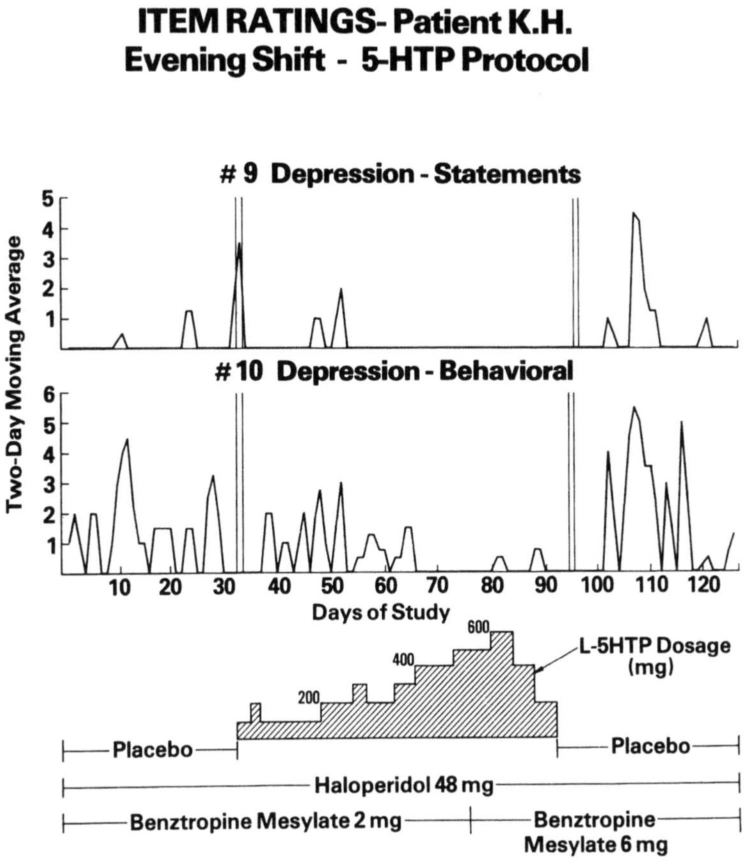
ITEM RATINGS- Patient K.H.
Evening Shift - 5-HTP Protocol
9 Depression - Statements
10 Depression - Behavioral
Two-Day Moving Average
Days of Study
10 20 30 40 50 60 70 80 90 100 110 120
200
400
600
L-5HTP Dosage (mg)
Placebo
Placebo
Haloperidol 48 mg
Benztropine Mesylate 2 mg
Benztropine Mesylate 6 mg

FIG. 18

a Brief Psychiatric Rating Scale (141) would have failed to record these fine changes. Clinically she presented a remarkable picture. Ordinarily when anyone would go on to the ward, she would assail them with a litany of delusional complaints; tales of the Mafia out to get her or her family or of bizarre creatures and devices out to harm her. Shortly after starting on the 5-hydroxytryptophan, she continued to seek out any newcomer to the unit but declared "Hey, doc, I'm getting better." This statement was accompanied by a warm smile and a motor restlessness bordering on a frank dyskinesia most disturbing to observe. As the dose of 5-HTP was increased, she became more alert, relevant and introspective. Delusional concerns decreased. She remarked that the voices from the sky had gone away. Her motor symptoms, however, continued to increase. She began protruding her tongue and licking her lips to an extent that made us fear she would cause some tissue breakdown. At the highest doses given she became mute and the increase in her bizarre behavior ratings overwhelmed the decrease in her hallucination rating to give an increased psychosis rating on the graph. All these changes are apparent only in the evening shift ratings, suggesting that for this patient changes were only visible or manifest during this less stimulating part of the day. With placebo substitution, she lost her bizarre motor symptoms but regained her delusional world.

While on 5-HTP she had begun to plan on going home and resuming her life there, and as she settled back into her more routine way of life she asked often for what she, in fact incorrectly, had identified as the active "amino acid pill." Even though she could not give a coherent or accurate account of what had happened to her biochemically, she was clearly aware that something had happened and desperately wanted to hold on to the changes she had experienced and was willing to ignore side effects that, to the clinical staff, were most distressing. During her optimal period her mother spontaneously commented on her marked social improvement, finding her better than she had been in many years. The physical side effects were, of course, distressing to the mother also.

The final set of graphs from the 5-HTP experiment are from a 20 year old man, ill for the previous four and a half years. Fig. 19 shows that while on active medication, his psychosis improved markedly, but like the previous patient he showed apparent worsening as higher doses were given.

Fig. 20 documents the relative resistance to change of the Global Mental Illness item. Although there was an observed decrease in hallucinations and in overall psychosis in this patient, this was not reflected in the ratings given for overall illness. We have found in many patients this question is most insensitive to change.

At this point you might well ask why you have been subjected to such a detailed exposition of the consequences of giving L-5HTP, Carbi-

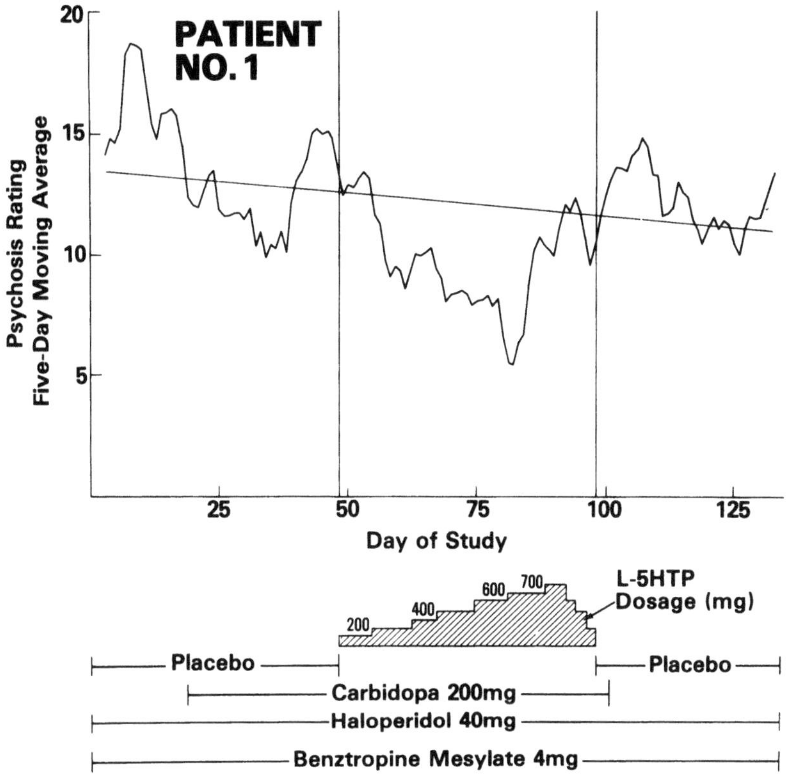
PATIENT NO. 1
Psychosis Rating
Five-Day Moving Average
20
15
10
5
25
50
75
100
125
Day of Study
200
400
600
700
L-5HTP Dosage (mg)
Placebo
Placebo
Carbidopa 200mg
Haloperidol 40mg
Benztropine Mesylate 4mg

FIG. 19

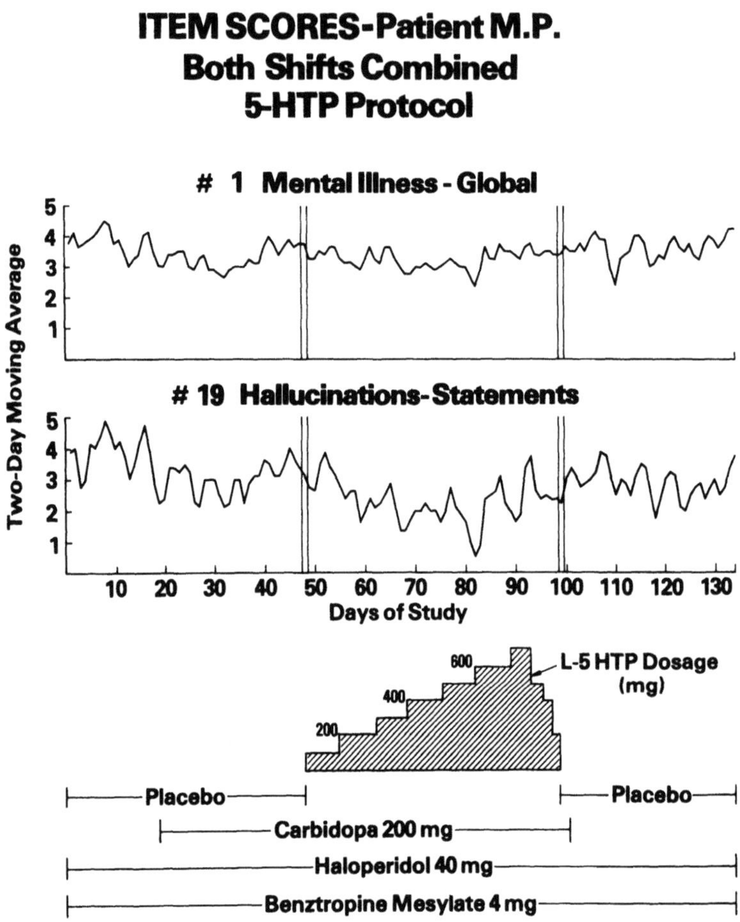
ITEM SCORES-Patient M.P.
Both Shifts Combined
5-HTP Protocol
1 Mental Illness - Global
19 Hallucinations-Statements
Two-Day Moving Average
5
4
3
2
1
10 20 30 40 50 60 70 80 90 100 110 120 130
Days of Study
600
400
200
L-5 HTP Dosage (mg)
Placebo
Placebo
Carbidopa 200 mg
Haloperidol 40 mg
Benztropine Mesylate 4 mg

FIG. 20

dopa, and Haloperidol to four chronic schizophrenic patients. If you view the experiment as a therapeutic trial of 5-HTP for schizophrenia you will conclude that it is another example of a therapeutic intervention in this syndrome producing a roughly even distribution of improvement, no change or deterioration. Thus we would have another typical bust in the annals of schizophrenic research, to be consigned to deserved obscurity and ambiguity.

I do not think that this would be a just fate. Indeed, these results highlight with great clarity some major issues that many may be aware of in this field, but few acknowledge sufficiently in their writing or research design. In an effort to avoid misleading others as well as in an effort to avoid being criticized as unscientific or even worse, anecdotal, most clinical biological researchers present only summed biological data derived from single point in time observations. The diverse results of this relatively simple experiment of giving a known biologically active precursor to patients carrying the diagnosis of schizophrenia argues strongly against such an approach. Schizophrenia really is a syndrome with diverse etiologies and our research protocols should begin reflecting this fact.

The data from the 5-HTP experiment and from the rating scale item correlations strongly support the contention that schizophrenia is a diverse syndrome and that therapeutic interventions should be designed so that individual case differences will be found and not buried beneath a mound of statistically averaged data. Close observation is essential, both over time and on a case by case basis.

Fig. 21 shows a segment of the 10 day moving average of the psychosis score for a patient not on any protocol. The large and short-lived dip in the score came as a surprise when the data were processed, and it is only in retrospect that it was established that this time period coincided with the patient's birthday. She had received a great deal of attention and visiting from her family and in this instance this had the effect of reducing the magnitude of her apparent illness to a striking degree.

How patients experience change is the next topic I would like to develop. I will present two patients involved in our reinvestigation of possible clinical effects of aqueous pineal extract. One patient appeared to derive substantial clinical benefit from the extract while the other showed statistically significant worsening (3).

First, the patient who appeared to improve when given extract. Fig. 22 shows the 10 day moving average of her psychosis score through baseline, extract, placebo and extract periods. There is clear indication of improvement and stabilization coincident with extract administration followed by relapse with placebo substitution. Fig. 23 shows the delusion and hallucination items from the rating scale fac-

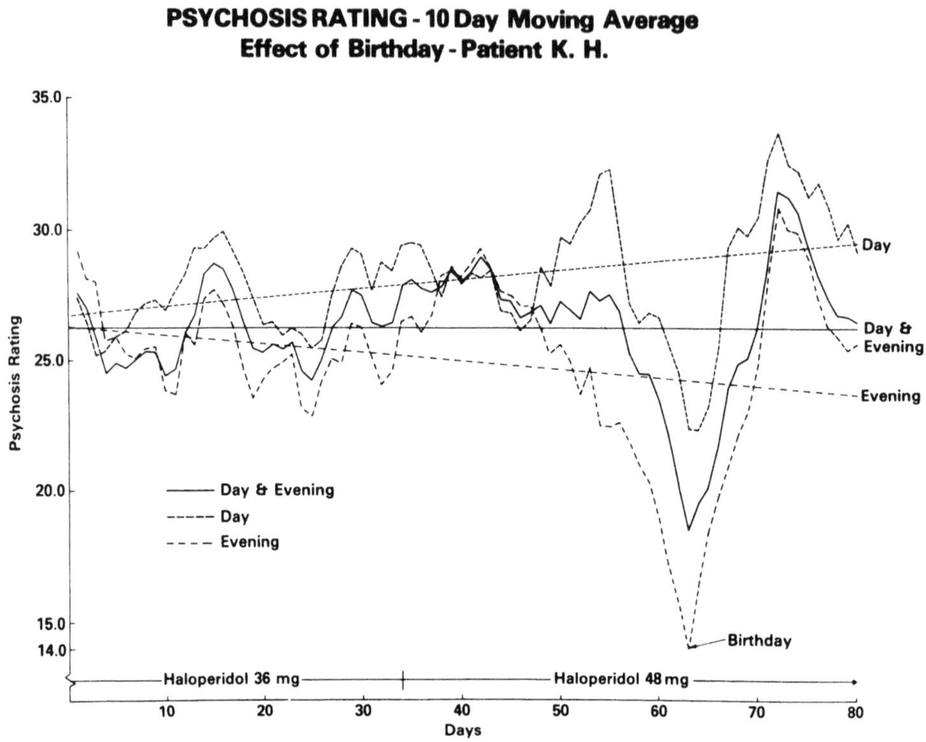
PSYCHOSIS RATING - 10 Day Moving Average
Effect of Birthday - Patient K. H.
35.0
30.0
25.0
20.0
15.0
14.0
Psychosis Rating
Day
Day & Evening
Evening
Day & Evening
Day
Evening
Birthday
Haloperidol 36 mg
Haloperidol 48 mg
10
20
30
40
50
60
70
80
Days

FIG. 21

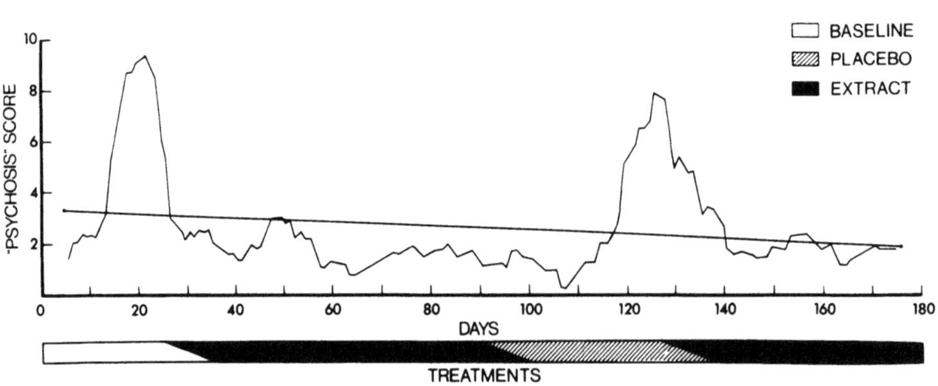
BASELINE
PLACEBO
EXTRACT
10
8
6
4
2
"PSYCHOSIS" SCORE
0
20
40
60
80
100
120
140
160
180
DAYS
TREATMENTS

FIG. 22

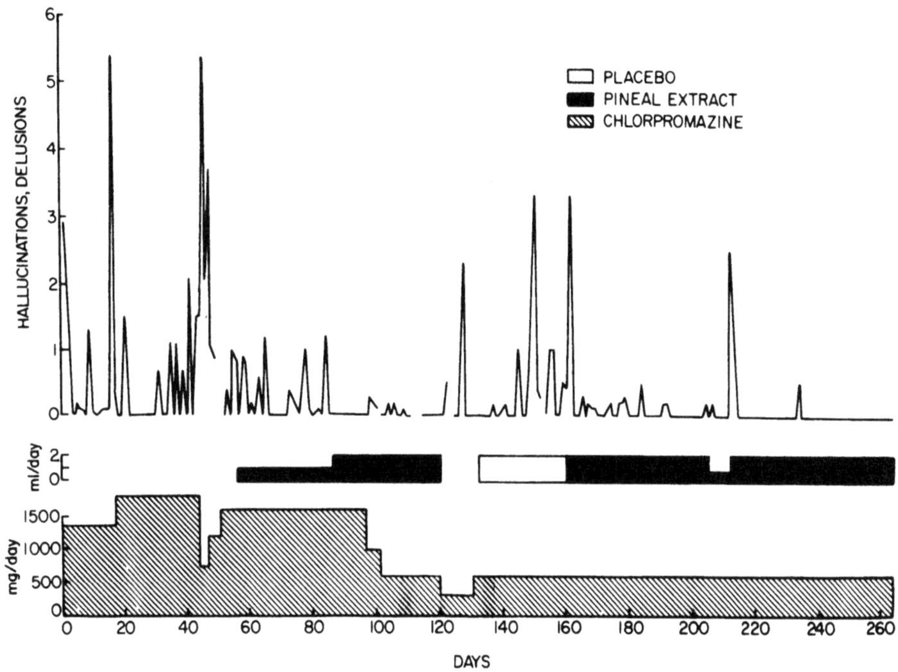
PLACEBO
PINEAL EXTRACT
CHLORPROMAZINE
HALLUCINATIONS, DELUSIONS
ml/day
mg/day
DAYS

FIG. 23

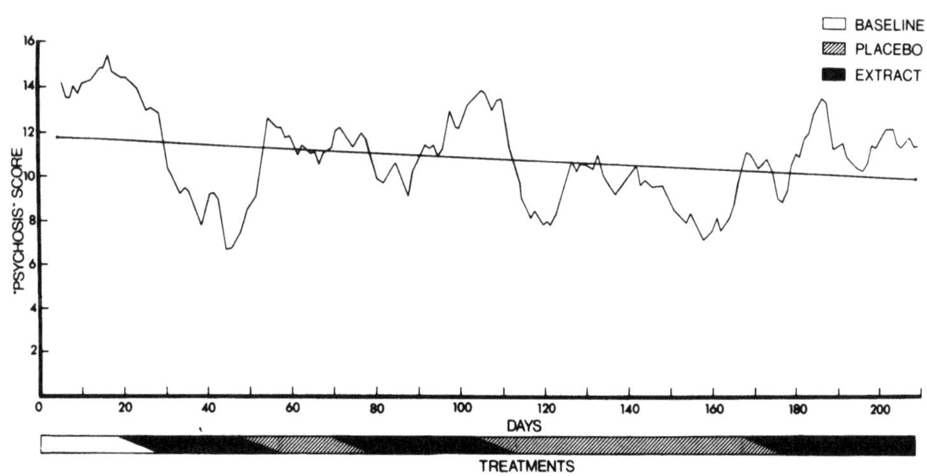
BASELINE
PLACEBO
EXTRACT
'PSYCHOSIS' SCORE
DAYS
TREATMENTS

FIG. 24

tored out and demonstrates similar improvement and stabilization coincident with administration of the pineal extract.

The patient was a 27 year old woman with a 13 year history, consistent with a "chronically acute" schizophrenic illness. She had spent fully eight of the past ten years in various hospitals. Her illness was characterized by extreme levels of anxiety, impulsiveness and fluctuating immersion in an hallucinatory experience. She was striking in her attempts to conceal both from herself and others the intensity and bizarreness of her experiences. Fig. 23 presents data for 260 days. At day 200 she was in a period of relative remission of her illness. Later her dose of extract was reduced -- with her knowledge -- and after a week, her psychosis erupted again. The clinical deterioration was obvious. She stated clearly, "You've done something to me." A blind crossover experiment done two months later showed similar changes with the exception that she was generally able while on placebo, to conceal the psychotic portions of her experience. She would only acknowledge grudgingly that "they are arguing a little," referring to her voices that stated alternately "We love you" or "We hate you." A subsequent extract period appeared to restore her ability to communicate more clearly and directly.

This particularly articulate patient is telling us things that I suspect others experience but are less able to communicate with such clarity. Research protocols and placebo substitutions have effects beyond those seen in our rating scales. These effects are difficult to quantitate, but that does not make them any the less important. Ignoring them means loss of important data and possible erroneous conclusions. It is my contention that, however ill many schizophrenic patients may seem, they are aware when we manipulate their biochemical inventory and may respond in ways hard to understand unless this awareness is taken into account.

The last clinical vignette I would like to present is from our experience with the patient in the pineal study who statistically and clinically appeared to show clinical improvement when first placed on extract, but who later showed little change and then marked deterioration on her subsequent two courses of extract. Fig. 24 shows the 10 day moving average of her psychosis score over 210 days. A short excerpt from her baseline period conveys a picture of a rather silly hebephrenic woman with a marked thought disorder. It is impossible to know if she is hallucinating or not. Neither can we tell if she has any formed delusions underlying her disturbed communication.

During a second course of extract treatment, we are given a glimpse into the underlying sadness this patient has kept from view. There is minimal evidence of schizophrenia, but her depression is clear. Her schizophrenia was present "with a vengeance." I tended to feel that she was making a mockery of me, the interviewer. There seems an almost willful element to her psychotic productions. Such is speculation, of course; impossible to prove or disprove. I submit that it is not unreasonable and failing to consider such possibilities can lead investigators to lose possible meaningful data or misinterpret the significance of their observations.

To summarize, I have presented an outline of a technique for observing and studying schizophrenic patients. Data from two protocols, L-5-hydroxytryptophan and aqueous pineal extract have been presented, not to argue that either is an answer to schizophrenia but as examples of biological experiments whose results cannot be comprehensively summed as indicating any single directional effect on the varied manifestations of what we call schizophrenia. I am arguing that progress in understanding and treating schizophrenic patients can come from individual case study. Indeed, such an approach really follows from the growing appreciation that schizophrenia is a syndrome not a disease.

I have further argued that we cannot regard our patients as biochemical variety stores simply in need of more efficient or better management. Experimental interventions have psychological consequences beyond their immediate metabolic effects, and although it is difficult to predict and allow for them ahead of time in a research design, they must be watched for and taken into account.

Both biological protocols clearly require considerably more work. The 5-HTP has to be dissected as to dose response and individual case response. Who is likely to respond and why? Such work takes an enormous amount of time and effort but will, I believe, pay off in the long run. Similarly where should the pineal work go from here? I think it probably should stop until likely biological chemical candidates are fractioned out for further testing. The work to date should be remembered and when new insights emerge as to substances possibly active in affecting parameters of mental function, they may be sought for in pineal tissue. We currently are trying to isolate a component of pineal extract that inhibits the in vitro formation of dimethyltryptamine.

DR. WOLF: I should think that initially it might be wise without being blinded to ascertain what mixture is going to work favorably with an individual. You have implied that these individuals respond characteristically to the medications whether they get better or worse, or do not change. That would suggest a strategy whereby you could tailor a therapeutic regimen until an individual is getting better and then blind yourself and replicate the process blind.

DR. BIGELOW: Right. It seems to me that a design of that

kind has the greatest potential for yielding the most useful information, particularly in small pilot studies involving relatively few patients. A preset fixed dose study of a research medication runs the risk that you will give too much or too little of it. Too much and the blind will be given away by the emergence of side effects, not to mention the possible vitiation of therapeutic benefit. Too little of an agent may lead to a false conclusion that the substance in that patient is inactive. To my mind the ideal study consists of individual titration of dose by a non-blind investigator followed by blind placebo substitution and later crossover back to active medication.

Hypothalamic Polypeptides

DR. LIPTON: We have been interested in the hypothalamic polypeptides. The old belief that hypothalamic polypeptides have one function and one function only, which is to migrate down to the pituitary and stimulate the pituitary into action, is probably not correct. They have other functions within the central nervous system. And there are data to support the notion that hypothalamic polypeptides are not exclusively in the hypothalamus. They can be found in all parts of the brain. How they get there, where they are synthesized, whether they do or do not migrate, one does not know. They also have behavioral effects. The experiment that I find most impressive is the one that Sam McCann and Pfaff did independently. In this they take a young female animal, hypophysectomize and ovariectomize it. They then allow it to grow and it becomes a sexual neuter behaviorally (129, 143). They can induce sexual behavior in that animal by giving it estrogens and then progesterone; this has been known and done for a long time. That is worth thinking about in itself. It involves inducing a complex pattern of behavior, assuming the position, receiving the male and possibly even attracting the male. All these are central nervous system activities. But if the animal is primed with estrogen and given luteinizing releasing hormone, not progesterone but LRH, the hypothalamic polypeptide, that whole pattern of behavior appears in an animal that does not have an ovary or a pituitary. What is involved with that behavior? It has obviously got to be that transmitters are involved in sensory perception, in the motor activity, in assuming the position, etc.

DR. WOLF: What you are saying is that a hypothalamic polypeptide that we identify as active on the pituitary may exert its effect systemically.

DR. LIPTON: In addition to effects on the pituitary.

DR. WOLF: This is, of course, very reminiscent of the somatostatin story. We know that somatostatin will inhibit both alpha and beta cells in the pancreas, for example, but people are very suspicious of it because of its locale and that it really ought to be operating on the pituitary.

DR. LIPTON: We have reviewed the data on the behavioral effects of the hypothalamic polypeptides and thyrotropin releasing hormone (TRH) will induce behavioral effects in animals and man. Drs. Prange and Wilson at our University have observed transitory improvement in schizophrenics after .5 mg. TRH was given intravenously. They are working with young, fairly active schizophrenics who are not on chlorpromazine. They are drug free.

DR. BIGELOW: I gather that their patients tend to be recent readmissions from the community; ours are not recent readmissions. Ours have not been able to maintain themselves outside of hospitals.

DR. LIPTON: Within the hospital Prange's patients are drug free for about a month. He gives them a single shot of TRH which is 0.5 mg. of TRH. He says he gets a partial remission, like a 50% remission very quickly, usually within less than 24 hours but lasting in some cases for days. There is some disagreement with that. They have made claims with respect to depression and again some agree and others do not. Some find it, and some do not find it. I think the important point here is that Prange does not claim to have a therapeutically useful drug in the sense that it is the break we have all been looking for in schizophrenia. What is important is that TRH has biological activities that alter the state of the schizophrenic very quickly, much more quickly than the usual pharmacological agents. As an investigative tool for understanding mechanisms, I find it very stimulating. One of the questions is, why do drugs used in the treatment of mental illness take so long to act?

DR. BIGELOW: I might add that our experience with TRH has been equivocal. We have run three patients. They were chronic but we could not see anything noteworthy and I wonder if Prange was dealing with a different clinical group. And I wonder if he found a few patients particularly responsive. Has he done tests, retests on those patients to try to confirm the phenomenon that he observed? It should be possible because the improvement is such a short lived phenomenon (189).

DR. LIPTON: Yes he should. He has done very few retests in chronic schizophrenics. The effect is not so short lived. In some cases it lasted 24 hours and a new clinical equilibrium was thereby established that continued for two or three weeks. In a few cases he was able to send the patient home without drugs.

DR. WOLF: Are you suggesting that one should distinguish drug responsive individuals from the drug unresponsive in the various categories including TRH and the more conventionally used agents?

Measurement of Drug Levels in the Blood

We have not said much about unresponsive people as distinct from responsive people. We should have one word in here about measurement of drug levels, how useful is that and how significant?

DR. GJESSING: Folke Sjoquist in Sweden has worked with these problems for a long time and he has shown that there is a difference in the percent absorption and metabolism of several different kinds of drugs, especially antidepressive agents but also others. They achieve very different levels in the blood on the same dosage, and the desirable level can be obtained by adjusting the dosages.

DR. WOLF: Can some drug nonresponsive people be made responsive by vastly increasing the dosage?

DR. GJESSING: Yes. To a level where therapeutic effects can be observed.

DR. SNYDER: I have always been struck by the enormous variability in the phenothiazine dosage requirement of schizophrenic patients. When the patient is titrated by a therapeutic response and you give them what is called the optimal dose, I am a little concerned about our skill in determining this. It looks as if before long there will be radio-immuno assays for phenothiazine. I have the feeling that this test will then become mandatory just as it is mandatory to do a hematocrit on a patient coming into the hospital.

DR. LIPTON: I do not think there is any question but that the clinical pharmacologist will note substantial variance in levels at a given dose when he is able to monitor blood level, whether it be with digitalis or psychotropic drugs. But I also think that patients with identical blood levels would still show vast differences in clinical response.

DR. ALTSCHULE: You cannot correlate the clinical response with levels of any agent. Not at all. Not in antibiotics, not in digitalis, not in anything.

DR. LIPTON: I think we would achieve an increased level of sophistication in dealing with the idiosyncratic patients if we could do blood levels and if these could be done easily. One more blood sample does not break the back of the laboratory, or the back of Medicare, or the back of the patient. And it would be worth doing. I am interested in variance. A thirty-fold variance in the imipramine levels after a standard dose strikes me as remarkable. One might accept two or three-fold variance, but a 30-fold variance is more than one expects. Perhaps great variance is one of the characteristics of schizophrenia or of psychosis, while normal subjects are more homeostatically stable. A markedly unstable internal milieu is incompatible with life. A slightly less unstable one might be incompatible with sanity.

DR. SNYDER: But blood level variations are as great in normals as in mentally ill patients!

DR. ALTSCHULE: One cannot draw inferences as to tissue levels from a knowledge of the blood concentration of a drug. Nevertheless this is as close as you can get.

DR. GJESSING: Dr. Altschule is correct. We would prefer to have the level in the nerve cell in order to get even better adjustments. But we cannot get much closer than the blood level.

DR. ALTSCHULE: Blood level does tell you one thing. It will tell you the degree of absorption. There is no question about that. If you have a drug that is not being absorbed, it will not show up in the blood.

DR. GJESSING: You can also study the rate of destruction or elimination of the drug.

DR. WYATT: With the tricyclics we can begin to see there are certain levels at which the anticholinergic effects are discernible and where you want to back off. You can see it clinically but you have confirmatory evidence in the blood level.

DR. LIPTON: All I am really saying is that one has the right at least to be suspicious. It does not tell you necessarily what is going on in the head, but if you find 30-fold variance in the blood you may very well get 30-fold variance in the head as well. And one at least ought to pay attention to that incredible degree of variance.

CHAPTER VII - Etiological Considerations

Participants: Drs. Bigelow, Bird, Cancro, Cohen, Gjessing, Goldstein, Kety, Lipton, Werthessen, Wolf, Wyatt

DR. GOLDSTEIN: In several neurological and autoimmune diseases such as multiple sclerosis and systemic lupus erythematosus, it now appears that immunological components and/or latent or defective viruses may be implicated. In looking at the biology and chemistry of schizophrenia, the question of a viral etiology should be reopened in light of newly developed concepts and tools in immunology. Perhaps Dr. Cohen, who has looked into this question quite closely, can give us some relevant information.

The Possibility of Viral Etiology

DR. COHEN: A hypothesis that invokes a viral etiology for schizophrenia should be able to account for the genetic patterns seen in chronic forms of schizophrenia. There are a number of animal models for which it has been shown that susceptibility of an organism to a specific virus is under genetic control. Such a mechanism may simply involve how the immune system of that organism handles the virus challenge. Craighead and associates have developed an intriguing model which demonstrates genetic susceptibility to a virus-induced diabetes mellitus-like disease in adult mice. The virus, a variant of the encephalomyocarditis virus, replicates in the pancreas and preferentially damages the beta cells in the islet of Langerhans. This virus can induce the disease in mice of one inbred strain, but not in those belonging to another strain. The very selective locus of extensive damage to the beta, but not to the acinar cells of the pancreas, is reminiscent of some mammalian viruses found in the central nervous system which may prefer neurons to glial cells, or may localize only in specific regions of the brain.

We also know of mammalian viruses which are primarily or totally endogenous. That is, they are not usually transmitted post-birth, but rather, in utero via the sperm, or the ovum, or some other mechanism. Recall that viruses are actually genetic vectors, sort of maverick chromosomes that move from cell to cell. Some even have the capacity to physically link their genomic DNA to the host cell chromosomal DNA and thus achieve a permanent, potentially functional state in the host cell and its descendents. It is entirely possible for a cell to harbor viral genes without any pathological signs or obvious metabolic evidence of the virus's persistence.

Regardless of whether a virus is transmitted endogenously or exogenously, susceptibility of the organism to the agent can be under genetic control.

DR. LIPTON: A virus ought to follow Koch's Postulates.

DR. COHEN: An editorial on "Koch's Postulates and Slow Infections of the Nervous System" by R. T. Johnson and C. T. Gibbs appears in the January 1974 issue of the Archives of Neurology, and specifically deals with the issues you bring up. These authors discuss Koch's postulates within the context of Koch's time and of our contemporary knowledge of viruses. They argue quite convincingly that, for a number of reasons, the literal content of Koch's postulates does not seem applicable to the attempt to establish a viral etiology for certain chronic diseases in humans. For example, Koch required that the parasite, after being isolated from the body, be grown repeatedly in pure culture without loss of activity. No viruses can be grown in pure culture on artificial media. A number of other recent studies which uncover the great diversity among virus replicative cycles makes literal application of the postulates equally untenable. In keeping with the spirit of Koch's postulates, however, Johnson and Gibbs do propose three tentative criteria which, if met, they feel would provide acceptable evidence to associate a specific virus with a specific disease process. In applying these criteria, the authors conclude that reasonable evidence has been accumulated for relating viral agents to four chronic neurological diseases in man -- subacute sclerosing panencephalitis (SSPE), progressive multifocal leukoencephalopathy (PML), kuru, and Creutzfeldt-Jakob disease. With my understanding of contemporary virology, I would conclude that we ought not necessarily require Koch's postulates to be met literally to prove a viral etiology for certain human diseases.

Koch's Postulates in Relation to Virus Infections

DR. WYATT: I have two questions. With the diseases where a viral origin is postulated, like multiple sclerosis or Jakob-Creutzfeldt disease, incidences do vary with geographic locations. The evidence we have about schizophrenia does not indicate geographic variability in incidence.

DR. COHEN: I think that it is not really clear for schizophrenia. There are many problems inherent in such epidemiological studies. E. F. Torrey and colleagues have been major proponents for the hypothesis that slow or latent viruses are involved in the etiology of schizophrenia. They recently reported a preliminary study on the epidemiology of schizophrenia in Papua, New Guinea, and concluded that there is a regional cluster phenomenon which cannot be explained away entirely by the possible weaknesses or biases in such a study. There are also reports in the literature of seasonal variations in that a disproportionate number of schizophrenic individuals are born during the first three months of the year.

DR. WYATT: The other question is, are there any suggested viral diseases now, other than perhaps schizophrenia, where there is not clear-cut anatomical destruction of some sort?

DR. COHEN: Yes.

DR. WYATT: Which? Are there any diseases that we know about in which viruses are involved without giving any evidence of their presence on microscopic examination?

Viral Effects without Destruction of Cells

DR. COHEN: R. T. Johnson studied mumps virus encephalitis in suckling hamsters (97) and rabies infection in mice (96). It was clear that clinical disease could result from a direct effect of the virus on neurons without detectable morphological changes in the infected cells. Only after use of fluorescent-tagged antibodies could large quantities of viral antigens be visualized within the otherwise normal-appearing neurons.

It should be noted that many of the viruses which cause lytic infections may also cause non-cytocidal infections, and some viruses induce only non-cytocidal infections. A cell infected by such viruses may or may not begin to chronically shed infectious virus or incomplete "defective" viral particles. A "producer" may not look or behave differently from an uninfected cell counterpart at the light microscope level. At the other extreme, after infection by a non-cytocidal virus, the cell may totally change its characteristics (e.g. transformation into malignant state), and yet there will be no detectable viral particles in the cell. We now know that the virus in this situation has actually inserted its genome into the host cell chromosome. It is thought that at least some viral genes are activated when the cell is in this transformed state. Also, we now have techniques to detect viral nucleic acid in such cells as well as to "rescue" from them virus particles by various treatments or manipulations of the cell culture conditions. A major effort has been directed towards trying to rescue virus particles from human tumors which usually show no evidence of virus particles. Interestingly, Meinke at the Scripps Clinic is now searching for evidence of viral genes sequestered among cellular nucleic acids from the central nervous system of patients with multiple sclerosis, amyotrophic lateral sclerosis and other central nervous system diseases.

DR. LIPTON: Do you then throw out Koch's Postulates?

DR. COHEN: Yes, some of them.

DR. LIPTON: What is the nature of the evidence that MS may be a slow virus or that the illness, like Kuru in New Guinea, is due to a slow virus? What kind of strategies are used when Koch's Postulates are discarded?

DR. COHEN: With respect to the kinds of strategies used, these are outlined by Johnson and Gibbs in their article to which I referred on page 136. First, it must be demonstrated in more than one laboratory that the disease can be consistently transmitted to experimental animals, or that a particular virus can be recovered consistently from the diseased tissues cultured in vitro. Second, it must be shown that the disease can be serially transmitted to animals with ultrafiltered material from the diseased individual. It must also be possible to transmit the disease with serial dilutions of the ultrafiltered material to establish that the infectious agent replicates. Alternatively, if a virus can be visualized initially, it should be possible to consistently find evidence of the virus (by electron microscopy, immuno-fluorescence, or other means) in the cells thought responsible for the clinical lesion. Third, it should be established that the agent in question is not ubiquitous or simply being carried along as a passenger or contaminant. It should not be found in normal tissues or in tissues from patients with other diseases. At the present time the studies on multiple sclerosis do not meet these criteria. On the other hand, Johnson and Gibbs feel these criteria have been met for kuru, despite the fact that it has not been possible thus far to demonstrate cytopathological effects in cell cultures, to visualize virus-like particles in infectious tissue or cell cultures, or to find evidence for an immune response in man or in inoculated laboratory animals.

Viral Effects on Immunity

Perhaps we should consider here persistent or recurrent virus infections which may influence brain function indirectly through effects on the body's immune system. In a recent publication (116) Robert Lewis discusses a very interesting series of slowly progressive multisystem diseases in animals which may provide models for spontaneous auto-immune diseases in humans. Interestingly, some of these animal diseases were at first thought to be of genetic origin; then they were classified as auto-immune diseases, and finally, they were shown to have a viral etiology. It appears that in these diseases, the virus sets in motion a mechanism for tissue damage via localized deposition of viral antigen-antibody complexes, so-called "immune complexes." Although many autoantibodies may arise following viral invasion, they may or may not be directly involved in the destruction of specific tissues and cell types. Systemic lupus erythematosus (SLE) in dogs and certain inbred strains of mice seems to be a virus-related disease of this nature. In a recent article (140), M. Oldstone discusses the fact that about 75% of humans with SLE have nervous and mental disorders for which neuropathologic correlations are often difficult to establish. Oldstone reports that he examined autopsy material from four patients with SLE with neurologic disorders and found immune complex deposits in the choroid plexus. Oldstone suggests a mechanism whereby immune complex deposition in the choroid plexus might disturb the blood-brain barrier and bring about changes in the brain ventricular fluids such as shifts in the pH and electrolyte balance. The glomeruli of the kidney and the choroid plexus act as sieves to remove circulating immune complexes. These immune complexes can interact with and activate the blood's complement system. The sequential activation of the various complement components by the immune complex releases a number of biological agents or activities such as immune adherence factor, anaphylatoxins, and vasoactive amines which ultimately induce cell membrane lesions, cytotoxicity and lysis. Fluorescent tagged reagents which detect immune complexes (i.e., tagged antibodies against human Ig, viral antigen, and C3, a complement component) are available. Perhaps we should begin to look for such deposits in autopsy material from schizophrenic patients or in biopsied kidney material obtained for other reasons from individuals with psychiatric disorders. It may also be worthwhile to look for anticomplementary factors in the Ig fraction of sera from patients with psychiatric disorders. It is interesting to note that psychiatric disorders sometimes appear in patients with Australia antigen hepatitis, a disease associated with immune complex formation. Perhaps we should search for antibodies specific for viruses most commonly found in the central nervous system such as measles, herpes simplex I, rubella and parainfluenza. Although much reduced in concentration compared to that found in the kidneys, such antibodies might be found to circulate as immune complexes in the serum or CSF. Indeed, there is one recent report in the literature of increased levels of neutralizing antibodies to herpes simplex among patients institutionalized with psychotic depression.

DR. LIPTON: These are plausible, exciting and very astute speculations. Could we hear more about the possible involvement of auto-immune mechanisms in schizophrenia?

DR. GOLDSTEIN: First we should define the three major cell populations that are involved in the conditions that favor auto-immune processes and immune complex diseases.

The Pathogenesis of Autoimmune Mechanisms

There are three major lymphocyte populations: the T-cells (thymus dependent) which are the effector cells which directly participate in cell mediated reactions, null cells which are potential T cells but are presently inactive and the B cells, which produce circulating antibodies. There is also a recently defined subpopulation of T cells termed suppressor or regulator T cells which seem to exert fine control over activities of both T and B cell populations. The disappearance of the suppressor T cells seems to correlate with the appearance of elevated immuno-globulin synthesis directed against certain cellular and humoral antigens. In recent studies of animals and humans which develop auto-immune disorders, it has been found that suppressor cells disappear just prior to disease onset. Without suppressor T cell control, other T cell populations apparently become auto-aggressive, auto-antibodies are produced and there may be an increased incidence of malignancy and early death.

DR. WOLF: It appears that a genetic susceptibility to a viral infection produces the pathological state. There are several models of this. Is it not also true that this pathological state can be accelerated or made more likely to happen, by environmental stress?

DR. GOLDSTEIN: Yes.

DR. WOLF: Dr. Goldstein and his collaborators at the University of Texas Medical Branch have been pioneering in the study of the hormone thymosin and its relationship to thymic lymphocytes as pertinent to the mechanisms of auto-immune disease. He has even been looking into the possibility that schizophrenia may fall in the category of auto-immune disorders.

DR. GOLDSTEIN: The ratio of the three major types of lymphocytes circulating in the blood (T-cells, B-cells and null cells) appear to change in specific auto-immune diseases. In a normal individual, roughly 20% of the blood lymphocytes are B cells, 65% are T cells and the remaining 10 or 15% are null cells. In certain auto-immune diseases the percentage of each cell type in the blood may vary significantly. For example, in patients with active SLE B cells in the blood comprise about 15% of the total lymphocyte pool. The percentage of T cells are reduced to about 40% and the null cells may be as high as 45%. In contrast, the percentage of B cells, T cells and null cells return to normal levels during the inactive form of SLE. Thymosin is a hormone isolated from the thymus gland which we have shown to enhance cellular immune function and increase the number of T cells, recognizable morphologically as E-rosettes (a lymphocyte that binds three or more sheep erythrocytes) in humans with cancer or with certain immunodeficiency diseases.

In collaboration with Drs. Scheinberg and Cathcart of Boston University College of Medicine, we have been looking at the effects of thymosin on the number of T cells in vitro from patients with active SLE. We found that thymosin significantly increased the number of T cells in the blood from SLE patients apparently by conversion of null cells into T cells. These studies suggest that thymosin levels may be deficient in patients with active SLE and as a result there is an accumulation of null cells (stem cells) in the blood. We routinely use this E-rosette plus thymosin test to screen for candidates for thymosin therapy. The E-rosette assay is diagramed in Fig. 25.

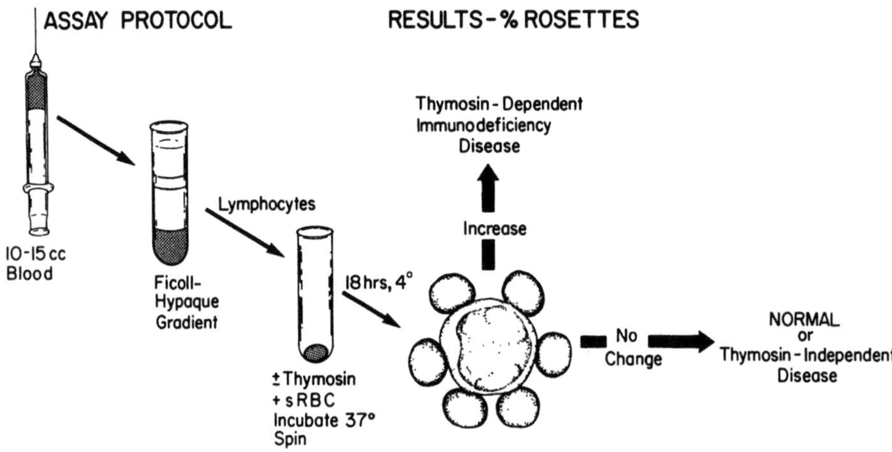

FIG. 25

An example of how thymosin increases in vitro the number of E-rosettes from patients with a variety of immunodeficiency diseases is shown in Table XV. In general, E-rosettes from "normal" individuals infrequently increase with thymosin incubation. Two exceptions we are aware of are lymphocytes from individuals with a history of allergy, or from individuals who are coming down with viral infections (colds). In the former case, we feel there may be a defect in the suppressor T cells which may be conducive to the development of the allergy, and in the latter case, the viral infection may somehow interfere with the production of thymosin by the thymus resulting in the accumulation of circulating null cells (stem cells). Both hypotheses are being investigated.

For the past few months we have been studying the lymphocytes of patients referred to Psychiatry at the University of Texas Medical Branch. To date we have examined in vitro lymphocyte function of over 40 patients with mental disorders including schizophrenia,

TABLE XV. Effect of thymosin (fraction 5) on human T-cell rosettes of immunodeficient patients

Published Reports	Baseline	Thymosin Treatment	Significant Increase
Wara & Ammann[1]			
Normals (20)	50-65%	45-65%	-
Thymic hypoplasia	20%	45%	+
DiGeorge syndrome	25%	49%	+
Thymic hypoplasia and successful thymus transplant	32%	38%	-
Ataxia telangiectasia	27%	43%	+
Wiskott-Aldrich syndrome	28%	45%	+
Combined immunodeficiency	17%	20%	-
Sakai et al.[2]			
Normals (8)	34-71%	43-72%	-
Lung carcinoma (undifferentiated adenocarcinoma)	24%	42%	+
Hodgkin's disease, IV-B	34%	47%	+
Acute lymphocytic leukemia	35%	47%	+
Acute myelomonocytic leukemia	23%	54%	+
Disseminated melanoma	23%	37%	+
Stem cell leukemia (complete remission)	71%	74%	-
Endometrial carcinoma (no metastasis found)	62%	70%	-
Acute myelogenous leukemia	2%	17%	+
Scheinberg and Cathcart[3]			
Normals (8)	58-67%	58-69%	-
Active systemic lupus erythematosis (6)	23-47%	35-72%	+
Inactive systemic lupus erythematosis (5)	52-70%	59-71%	-
Rheumatoid arthritis (5)	60-69%	58-67%	-
Thurman and Goldstein[4]			
Normals (26)	52-81%	42-76%	-
Allergy patients (6)	48-64%	59-74%	+

1. Wara, D.W., Goldstein, A.L., Doyle, W. and Ammann, A.J. New Eng. J. Med., 292, 70-74, 1975.
2. Sakai, H., Costanzi, J.J., Loukas, D.F., Gagliano, R.G., Ritzmann, S.E. & Goldstein, A.L. Cancer 36: 974-976, 1975.
3. Scheinberg, M.A., Cathcart, E.S. & Goldstein, A.L., The Lancet, 424-429, February 22, 1975.
4. Thurman, G.B. and Goldstein, A.L. Bollettino dell' Instituto Sieroterapico Milanese 54, 203-210, 1975.

schizo-affective disorders, organic brain syndrome, character disorders, psychopathic personalities, hysterical psychosis, anxiety neurosis and reactive depression. Some of these patients are on drugs and others are not. Surprisingly a significant number of these patients, including the schizophrenics, have lymphocytes which respond to the hormone thymosin in the E-rosette test similar to that found with lymphocytes of highly allergic individuals or individuals with viral infections. It is a curious observation, and we are not sure yet what it means. We have looked also at other parameters of lymphocyte function in schizophrenics and as shown in Table XVI they appear to be normal. We have examined the capacity of T cells and B cells of schizophrenics to respond to mitogenic agents and have seen no significant change thus far. The E-rosette findings are interesting as we think this assay may measure a more immature population of T cells. We see significant increases not only with lymphocytes from schizophrenics but with lymphocytes from patients with a whole variety of other mental disorders. We are going to try to sort these out. We have yet to explore the role of stress, poor nutrition, psychoactive drugs and other environmental factors on the T-cell phenomenon we have observed.

Immunologic findings in Schizophrenia

DR. KETY: What is that column headed control? And why is there such a big difference?

DR. GOLDSTEIN: The "control" category is a baseline. It consists of cells cultured without any mitogenic agents. Due to the large amount of variation in the cultures not containing mitogens within the control and schizophrenic groups, there is also no significant difference between these two groups with respect to the cultures containing mitogens.

I might add that, in contrast to our studies, a group of Russian investigators under the direction of Dr. Marat Vartanian have found differences between schizophrenics and normals with respect to the responsiveness of lymphocytes to the mitogen, phytohemagglutinin (PHA) (181). They find a significant reduction in the response of lymphocytes from schizophrenics. These same investigators reported two other major observations which are of interest: (1) they find an antibody-like substance in the blood of schizophrenics which cross-reacts with mouse thymocytes, and (2) they find that lymphocytes from schizophrenic patients are much more fragile than lymphocytes from normal individuals. With regard to their first observation, we have attempted unsuccessfully to repeat their observations at the University of Texas Medical Branch with newly diagnosed schizophrenic patients. We did not find anything in the serum that lysed mouse thymocytes. In addition, a Canadian group (168) was also unable to repeat their findings. Nevertheless,

TABLE XVI

RESPONSES TO MITOGENIC STIMULATION OF LYMPHOCYTES FROM SCHIZOPHRENIC PATIENTS

	Uptake of ^{3}H-Tdr at 72 hours of culture (average counts per minute ± standard error)			
	Control	PHA-P	Con A	PWM
Normals	796 ± 334	99,702 ± 15,303	83,848 ± 14,280	28,715 ± 5,498
Schizophrenics	360 ± 38	88,224 ± 14,020	77,134 ± 14,434	47,889 ± 11,445
t-value	1.43	0.56	0.33	1.42
degrees of freedom	9	9	9	9
Difference ($p = 0.05$)	No	No	No	No

negative results are not definitively conclusive.

Studies in mice have clearly demonstrated that thymocytes and brain cells share several membrane antigens in common, and that an anti-mouse brain serum can lyse mouse thymocytes and vice versa. Most recently, Isaac Witz from Tel Aviv University and colleagues have found something more frequently in the serum of schizophrenics which combines with specific areas of the brain (191). Thus, it may be worthwhile to resurrect the hypothesis of anti-brain antibodies or factors as a cause of schizophrenia and to survey a greater number of patients.

The second interesting observation that the Russians have made deals with the fragility of specific populations of lymphocytes from schizophrenics. Using simple cytotoxicity techniques, they have evidence that the lymphocytes of schizophrenics are more fragile than those of normal subjects and as indicated above their response to mitogenic agents like phytohemagglutinin (PHA) is significantly reduced.

DR. LIPTON: Do you have any data on just what cortisone will do?

DR. GOLDSTEIN: We have no direct information regarding iatrogenic injections of cortisone. Some stressed individuals we have looked at appear to have normal lymphocytes as measured by the E-rosette bioassay. In the mouse, cortisone would knock out the more immature T cells.

Well, what has all of this have to do with the etiology of schizophrenia? At this point any relation we make is mostly conjecture. If we hypothesize that schizophrenia might be an auto-immune disease, what might we expect to see? In the NZB mouse, a mutant strain that develops auto-immune disease, there occurs a decrease in thymosin production by the thymus. Lower levels of thymosin in the blood would lead to a decrease in suppressor T-cells and an increase in thymosin-responsive cells in the blood (null cells). A decrease in suppressor or regulator T-cells might result in hyperactive B cells and in the formation of auto-antibodies and immune complexes. If there is an auto-immune component in some brain disorders and based upon in vitro thymosin studies we have completed in patients with known auto-immune diseases (e.g. SLE or Sjogren's disease) you might expect in schizophrenic patients to find an increase in null cells in the blood. If our very preliminary data are correct, we do indeed find a greater number of thymosin-responsive cells in the blood of schizophrenics. Much more extensive studies are required to establish a role for the immune system in schizophrenia and other mental disorders. Our studies to date have merely stimulated our interest in the possibility that auto-immunity and/or immune-complex deposition may be involved in schizophrenia. Further research in this area is planned.

DR. WOLF: Dr. Cancro mentioned animal models. Have any of you seen the pointers in Arkansas? Let me describe them to you.

Animal Models

Murphree, Dykman and Peters in the Department of Psychiatry at the University of Arkansas started breeding pointers a good many years ago. They found that they could inbreed strains of pointers that were normally friendly and that acted like any pointer friend that you or I might have, but another breed that was "squirrelly." This strain which separated from the normal pointers for successive generations was characterized by an implacable suspiciousness of people. It was impossible for anyone to get close to such a dog, to scratch the back of his neck. The dogs walked around looking apprehensive. If you took one of the dogs and held his paw, he would take a cataleptic posture. But they had a very difficult time getting support from the National Institute of Mental Health when project site visitors first started visiting their laboratory, because it was felt then that heredity had really nothing to do with behavioral disorders. This was about twelve years ago. The project site visitors looking at the two groups of dogs would say, "I understand you have been mistreating these dogs. They look like dogs that have been mistreated." The researchers denied this but the visitors went on, "The attendant there, the diener, comes in at night and beats these dogs up unbeknownst to you." To make a long story short, the investigators tried all of the maneuvers, cross-fostering and what not, and found that the behavior of the shy dogs as they called them, could not be modified by taking them immediately after birth and putting them with a bitch from the friendly group. By the same token, if they put a friendly dog in with the shy bitch, the friendly dog turned out friendly. They could enhance the friendliness of the friendly dogs by petting them when they were tiny little puppies, the way we do with all dogs we are breeding for experimental purposes. But you could not get anywhere with the so-called shy dogs. It was a very impressive study (131).

The investigators are able to predict the numbers of "shy" dogs in cross breeding maneuvers with the friendly ones. It seems to me that here is an example of something going to waste as it were, for lack of an interdisciplinary biochemical approach to what is happening. They have been doing some drug treatments of these dogs and actually have been able to benefit the dogs with phenothiazines. It seems to me it would be a golden opportunity for one of you with a sophisticated laboratory to collaborate with these men because they have a predictable model of a kind of behavior which is easily described which resembles and presents many of the anthropomorphic features that we think of in schizophrenia.

DR. BIRD: Among shy pointers the descriptive term is not "squirrelish" it is "wolfish"!!

DR. LIPTON: I have had some experience with animal models that goes back a long time. Unfortunately if you go to any veterinarian and ask if they have ever seen crazy dogs, they reply, "Why, of course, paranoid dogs, autistic dogs, all kinds of dogs." If you suggest that these dogs might be bred, they think you are out of your mind. They want to improve the breed. They kill these dogs off as quickly as they possibly can. We have been collaborating for some years now with the McKinney-Harlow group at Wisconsin. The genetics of the monkeys employed are not at all clear but they can certainly produce very strange animals with pitting and isolation techniques. These are environmental manipulations. That has interesting implications. They can produce animals with bizarre behavior which improves with chlorpromazine. Curiously they can also improve their behavior with what they euphemistically call psychotherapy, using monkey psychotherapists.

DR. WOLF: How about megavitamins?

DR. LIPTON: To the best of my knowledge they have not touched those. The psychotherapist is a baby monkey that, needing contact, persistently approaches the sick monkeys. The babies must be carefully protected, because the sick animals may be vicious and unwilling to tolerate approaches. After some time, however, the persistent baby monkey can make them socialize. Then they can enter a group and start to look like normal monkeys. Prange and Bresse at our school examined the brains of monkeys sacrificed during the period of their bizarre withdrawn behavior. Nothing turned up but perhaps we looked for the wrong things. We did find changes in the adrenals but not in the brains as such.

DR. WOLF: The monkeys you describe have induced behavioral disturbance, which is quite different from the hereditary mechanism. I just wanted to point that out.

DR. LIPTON: Yes. But though it is psychologically induced, it can be treated pharmacologically, with significant improvement. Or it can be treated by this so-called monkey therapy, using a monkey therapist. That is a psychological intervention which also yields beneficial results. I am impressed that both pharmacotherapy and psychotherapy are beneficial.

DR. WERTHESSEN: What happened to the adrenals?

DR. LIPTON: There is increase in adrenal tyrosine hydroxylase, for example. There is adrenal enlargement with increased cortisol and epinephrine production. Those were the primary changes we were able to find. In the brain we could not find anything. That was a disappointment.

Turning to another fascinating subject, Sam Corson in Ohio State has been experimenting with two categories of dogs that are hybrids with unknown genetics (34). Corson is a Pavlovian. He found dogs that he could not condition. If he tried to put them in a harness, they tore the harness apart. Those dogs, interestingly, do not respond to phenothiazines but they do respond to amphetamines. And not only do they respond to amphetamines, but when you give them amphetamines, you can put them in a harness, you can condition them and after they have been conditioned, you can take away the amphetamines and they will still tolerate the harness and will still maintain the conditioning. A third type is to me the most amazing. He has found some dogs so vicious that their owners are prepared to kill them. Food has to be thrown to them because no one dares to go near them. If those animals are given amphetamines, they also calm down, become friendly, can be petted, fed and played with. After this has gone on for some months you can stop the amphetamines and they remain friendly dogs. I do not know what goes on in there but it is a phenomenon that I think cannot be ignored in terms not only of pharmacological alterations in behavior, but of the additional fact that many of the changes induced while on the drug persist after the drug is discontinued. What happens to the brain I have no idea, but it is another example of drug-environment interaction.

DR. GJESSING: Is that not similar to the treatment of hyperactive children with stimulants (methylphenidate)?

DR. LIPTON: Yes, in a sense but usually such children must be treated for years. Unfortunately there are no studies in children that combine amphetamine use with psychological treatments to see whether they are additive or synergistic.

DR. BIGELOW: Yes, in children treatment has to be continued until puberty at least. Withdrawal of the drug will lead to a return to hyperkinetic behavior.

DR. CANCRO: After puberty withdrawal of the drug is not followed by return of the old condition. Dr. Ginsburg (65) at the University of Connecticut has been working with a strain of dogs which he is seeking to develop into an animal model of hyperkinesis. These dogs are a cross-breed of terriers, beagles and others. They found that there are two somewhat different strains of "overactive" dogs. This means that they did not learn inhibitory conditioning. They would not learn to sit or to stay. You would try to teach them to sit and they would sit but immediately they got up, walked over and wanted to play with you. And some of the animals responded to amphetamines and would learn this inhibitory conditioning, and other genetically different ones did not respond to amphetamines. They have a pretty nice array of responders and nonresponders which represent different mixtures. The model seems to be holding up at this point at least as an animal model. The conclusion is that there are potentially many useful models.

DR. LIPTON: There are two other models that I find interesting. None of them have all the features of schizophrenia. In some nutritional studies a sort of Kwashiorkor is induced in the monkey. Such monkeys can learn simple tasks. If the tasks become difficult, differences can be shown in learning between the normal monkey and the "retarded" monkey. The Kwashiorkor monkeys show strikingly a phenomenon that seems to be psychologically characteristic in schizophrenia. It is called "neophobia." The first time I ever heard that term used was by Cameron describing autistic children and later by Dave Shakow who used the term to describe the inability of schizophrenics to change the "set" in tasks. These monkeys show a neophobia. They have learned the task and then if you change the interval between the stimulus and response, change the nature of the task they are no longer able to perform.

Finally, there is a curious rat model, again malnutrition induced protein deficiency, in which there is not only diminished exploratory activity, but also stereotypy, freezing, bizarre types of behavior in young rats. This condition is also curiously responsive to the introduction of a "psychotherapist" rat. This is an adult, virgin female who cannot possibly give them milk but who grooms them, picks them up or is attentive in other ways. This treatment eliminates the bizarre behavior, even though the nutritional state is not improved. So there is interaction between nutrition and psychological stimulation. Such models which display certain features of an illness rather than the total illnesses may be useful in research.

DR. CANCRO: Such models can occur spontaneously or be produced by environmental manipulations. Jack Calhoun reported years ago "autistic" behavior in his mice colonies as a consequence of overcrowding (19). They would not mate, they lost interest in playing, they wandered around, they would not eat, they would not groom themselves, they did not socialize and their behavior was disorganized. The animal models have another interesting feature which I believe corresponds to the clinical reality in that the animals even achieved the same behavior through different biologic pathways. That strengthens the usefulness of these models because part of the clinical problem in the schizophrenic disorders is to disentangle what the pathways are, and to recognize that there are multiple pathways leading to the same manifestations.

DR. WOLF: One other model that should be mentioned for the

record, is the Horsley Gantt dog (58). This was another animal that was induced to develop this peculiar withdrawn behavior and would also demonstrate catalepsy. After conditioning the dog to do a task he confused the signals, forcing the dog to make increasingly difficult discriminations in a task where he was originally successful. And after a long period of such treatment this dog's behavior became permanently altered, closely resembling the behavior of the Murphree, Dykman, Peters dogs. If you stand in their corner, they will slink around the edge of the wall and go to another, and it is almost as frustrating to try to communicate with the dog as it is trying to communicate with a schizophrenic patient.

DR. LIPTON: These examples I think serve to emphasize that there are lots of ways in which the same peculiar type of behavior can be made to occur. There are also some very interesting interactive type studies reported in a publication by the Swedish Nutrition Council (35).

Frankova, a Czechoslovakian nutritionist, measures exploratory behavior in well-nourished rat pups, who are living in the ordinary rat pup social ambience (52). Normal behavior was designated as 100% (Fig. 26).

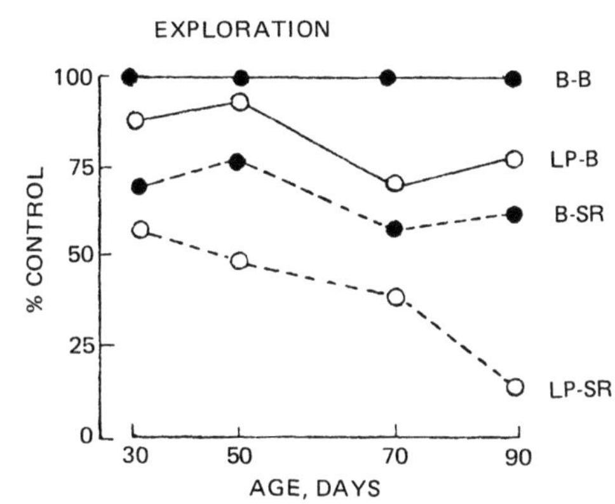

FIG. 26. Exploratory activity in adult rats raised in two different nutritional and sensory conditions. Nutritional treatment: diet containing 25 cal. % protein (B); diet containing 8 cal. % protein (LP). Sensory conditions: basal in animal rooms (B); restriction of optical and acoustical stimuli and contacts with experimenters (SR).

(Published with permission from S. Frankova: in Early Malnutrition and Mental Development. Symposia of the Swedish Nutrition Foundation XII (J. Carvioto, L. Hambraeus, B. Vahlquist, eds.) Almquist and Wiksell, Uppsala, 1974, pp. 202-210.)

The activity of rats on a low protein diet decreased to 80%. If, on the other hand, the well nourished rats were exposed to partial social deprivation or sensory deprivation, isolated from the litter with limited access to the mother, their behavior also decreased to about 80%. Social deprivation of the malnourished rats resulted in a decrease in behavior of about 30%. So if they are marginal in protein and socially deprived, behavior deteriorates greatly whereas exposure to only one of the deprivations at a time produced very limited change.

The process is reversible if the surrogate mother is brought back. Such studies establish that there may be interaction among various etiologic dimensions or parameters relevant to a syndrome or disease. Thus nutritional and psychological elements in the environment may interrelate causally.

CHAPTER VIII - Therapy

Participants: Drs. Altschule, Bigelow, Bliss, Cancro, Gjessing, Kety, Lipton, Snyder, Wolf, Wyatt.

Drug Effects

DR. KETY: It is my impression that the dosage of phenothiazine which produces a marked change in schizophrenic behavior does not affect normal behavior very much after the initial sedative effects of the chlorpromazine have yielded to tolerance. An initial dose of chlorpromazine will knock us all out but the normal individual very quickly develops tolerance to that. Then a normal person given the same dose as a schizophrenic would only show a very modest kind of retardation. Am I right about that?

DR. BIGELOW: I did not realize there had been chronic thorazine studies done with normal subjects.

DR. CANCRO: Well, I do not know of any chronic/normal thorazine studies but I do know that we have given a hundred or two hundred milligrams of chlorpromazine intramuscularly on admission to schizophrenic patients and it merely reduced their agitation. I am convinced that if I had given that dosage to myself or one of the residents, we would have been moribund. But I do not know.

DR. BLISS: It may be the agitation itself which increases tolerance in schizophrenia.

DR. WOLF: This is also true with sodium amytal and other barbiturates. You can give a catatonic or an excited schizophrenic enough to put a normal out for two or three days; the schizophrenic may not even go to sleep.

DR. KETY: While it is true that the dose of a barbiturate which will quiet a schizophrenic, will produce a marked hypnosis in a normal person, phenothiazine is more specific. It acts in schizophrenics in a way that is different from the way it acts on normals.

DR. WYATT: Phenothiazine has other effects. It affects blood pressure. If you take an excited chronic schizophrenic off phenothiazines for a long period of time, not for a couple of weeks but for several months, then upon resuming the drug they are hypersensitive. Their blood pressure is unobtainable and they become profoundly sedated, even if they were very excited. So there is a changed reaction. It is a complex phenomenon.

DR. GJESSING: I was thinking of the barbiturates. In 1922

Klasi in Zurich treated schizophrenics with sleep therapy. He kept 26 patients on high doses of barbiturates day and night over a week or more. One-third of these patients were very much improved or discharged, some were temporarily improved, and some not improved. The explanation of this phenomenon may be that this prolonged sleep precipitated a spontaneous remission.

DR. KETY: Indeed, the semi-narcosis treatment of schizophrenics with sodium amytal was a very dramatic phenomenon. I never forgot seeing this in a series of catatonic severely retarded patients. Fritz Frehan gave them sodium amytal and they came out of this state immediately and began to talk. However, in ten or fifteen minutes they would revert to their former state. One does not see that very often. But that convinced me 25 years ago that there must be some biological substratum in schizophrenia, else how could a chemical substance produce such a dramatic change so quickly.

DR. WYATT: Is that the implication? If the schizophrenic process could be reversed even for 15 minutes with the drug, does that mean that the person had been hopelessly irreclaimable before?

DR. GJESSING: Certainly.

DR. CANCRO: I think I would agree with the conclusion but I do not think I would agree with the logic of your argument. For example, when Dr. Kety was referring to the social withdrawal, we do not know if this is part of the process of the disease or if it is a way of the patient's coping with some aspect of the illness which is frightening to him. It is very hard to say how much of what we are seeing clinically at any given moment is part of the process or secondary to the process or a derivative consequence of the process. In a sense the amytal may be doing something that the patient is trying to do for himself through withdrawal. It may reduce the pressure so that temporarily he is more comfortable in talking with you. If, however, during those 15 minutes, there is no thought disorder, the delusions and hallucinations are gone, and the affect is good then I would say this would be powerful evidence in favor of a biological substratum. I think the phenothiazines have made such a remarkable difference because the patients get better whether you accept the fundamental symptoms of Bleuler's concept or not. They get better in ways that are substantially different from mere sedation. Their thinking gets better and their relatedness gets better, not to mention the improvement in their ability to function, to tolerate stress, and to meet the demands of day-to-day life. So we are doing something fundamental with the phenothiazine that we were not doing before, although we do not understand what it is.

DR. GJESSING: With sleep therapy, there were remissions that lasted for a long time. Some of you may remember that remissions lasting a week or two followed fever therapy. So there are many different kinds of possible manipulations.

DR. LIPTON: An appendectomy or a cholecystectomy would occasionally induce a remission for a week or two. I think it is an important issue to keep in mind.

DR. WYATT: From a survey of the literature on neuroleptic drugs we identified fifty that appear to be equally efficacious.

Comparative Efficacy of Drugs

Generally, the standard has been chlorpromazine which has been set traditionally at a hundred for comparison with the effects of other drugs. We have a relative potency then in terms of a comparison to chlorpromazine and how much drug do you need to get the same effect as you do with chlorpromazine. Thioridazine rated 108 and acetophenazine 22. We then tried to make a similar rank order for drugs used in depression, again setting chlorpromazine as 100. Unfortunately, there are not as many neuroleptic drugs which have been compared to chlorpromazine on depression but there are a sufficient number, so that you get some idea as shown on Table XVII.

The relative potencies come out very close with a product moment correlation of .86 which is, of course, significant. Then in a comparison of the effects of drugs in schizophrenia we picked oxyperitene as a standard, a neuroleptic drug which has been used only once in a comparative depression study. Again we obtained a very high correlation of relative potency. What is the significance of this? Perhaps depression and schizophrenia are the same disease since you have drugs that are equally efficacious in both, or alternatively, the drugs may work through relatively non-specific mechanisms, for example a dopamine system, that is not the primary cause of the illness but might be an intermediary for both depression and for schizophrenia.

DR. KETY: Dick, what symptoms of depression are these phenothiazines effective for? Are these papers talking about agitated depression, psychotic depression? If they are talking about psychotic depression, this could still be an anti-psychotic effect. If they are correcting the depression, that would be interesting.

DR. WYATT: I would like to be able to answer that question but I cannot. And that is because the depressionologists probably have not been as careful in defining what they are talking about as schizophrenologists who have been extraordinarily careless, as we all know.

TABLE XVII

RELATIVE POTENCY OF NEUROLEPTICS
IN SCHIZOPHRENIA AND AFFECTIVE DISORDERS

DRUG (AND CLASS)	RELATIVE POTENCY	
	Schizophrenia	Affective Disorders
CPZ (Phenothiazine, Aliphatic)	100	100
THIORIDAZINE (Phenothiazine, Piperidine)	108	128
CHLORPROTHIXENE (Thioxanthene)	35	42.5
ACETOPHENAZINE (Phenothiazine, Piperazine)	22	105
HALOPERIDOL (Butyrophenone)	2.4	2
PIMOZIDE (Diphenylbutylpiperidine)	1.2	4
MESORIDAZINE (Phenothiazine, Piperidine)	45	70
OXYPERTINE (Phenylpiperazine)	38	30
LOXAPINE (Dibenzoxazepine)	9.2	5
PERPHENAZINE (Phenothiazine, Piperazine)	8.8	15
THIOTHIZENE (Thioxanthene)	3.2	3

DR. CANCRO: We have an even worse situation, because people use the term depression to mean everything from a depressive mood which is ubiquitous, particularly in these times, to a depressive illness and will rarely specify what they are talking about. It is very much to the heart of the matter to ask about the quality and kind of control studies particularly about placebo effect. Placebo effect is a recurrent problem in the treatment of "depression."

DR. WYATT: I think we can argue against the placebo effect because you have comparable efficacy. Many of the neuroleptics were initially compared to one of the tricyclics. Most people would agree that these tricyclics do have an effect above and beyond placebo.

DR. LIPTON: You say you cannot do anything about symptoms. I would imagine most of the depression studies used a Hamilton scale or something like that. Cannot you extract out those symptoms that might conceivably be problems in both schizophrenia and depression?

DR. WYATT: We will give it a try. We could go back and do that.

DR. KETY: To get back to what Dr. Cancro was saying, we are also doing an affective disorder study like the schizophrenia study. I must say the diagnosis of schizophrenia was difficult but the diagnosis of affective disorder was just a terrible can of worms. The only affective disorders that I think one can really be pretty sure of, is clear cut manic depressive psychosis. But neurotic depression and unipolar depression is just a mess. Unless one took a very specific depressive syndrome, I would worry about it.

DR. SNYDER: Yes underlying the rationale of Table XVII is that anti-schizophrenic drugs are as good in depression as they are in schizophrenia or another way of looking at it would be that these drugs, phenothiazines and related drugs are as good in treating depression as traditional anti-depressants, and this is of course a major issue in psychopharmacology which has been addressed in a number of studies. Of course if that were the case the whole classification of drugs would become absurd. All of the studies were derived from original research. First there were studies showing that thioridazine was good for depression. Then someone said chlorpromazine was. And then there was a study with perphenazine. What does it all mean and how do you put it together? Leo Hollister feels, after analyzing extensive data, that when the phenothiazines or related drugs are used in depression, they are not really effective for depression in the way that tricyclics are. Tricyclics have their unique and major efficacy in retarded endogenous depression, and drugs like thioridazine and chlorpromazine seem to be more useful in anxious depression which Hollister feels is relatively non-specific and might be anticipated from some kind of general tranquilizing action of drugs, not necessarily sedating but not really a classic anti-depressant.

DR. CANCRO: You have this problem that Dr. Kety was talking about a moment ago. A person is restless and agitated and has an increase in motor activity. He does not seem to be able to control these symptoms and feels unhappy and depressed about it. Which is the noun, the depression or the restlessness? I do not think it is an accident that in the thioridazine study, at least clinically, the patients who respond are those who present with restlessness, motor agitation, anxiety, and who do not have the classical depressive picture of life is not worth living, everything is bleak, no desire to do anything, a loss of interest in everything, etc. In this classic endogenous recurrent depression, Hollister said that the phenothiazines are not efficacious.

DR. WOLF: They were tranquilizing without being anti-depressant.

DR. KETY: It revolves around the rating scale. What are the features in the Hamilton rating scale and are they given equal weight? I do not know too much about it but it seems to me that depression itself, the mood, the depressed mood, is probably only one small part of the depression scale. Restlessness and other features are common to schizophrenia.

DR. WYATT: The Hamilton scale covers many complaints: insomnia, weight loss, somatic complaints, agitation, symptoms that you identify with depression. It also covers depressed moods.

DR. GJESSING: You said that many of these were manic?

DR. WYATT: Only one of the studies had, exclusively, manics in it. That was the haloperidol study. Some of the rest of them had a mixture of mania and depression. Others had patients who were only depressed.

DR. CANCRO: It also depends on the year of the study. Up until the advent of lithium therapy, nobody was making the diagnosis especially in the east. It just did not exist. Fortunately we rediscovered it.

DR. WOLF: Rediscovered mania mania.

DR. KETY: Somebody said that lithium is being used effectively in the treatment of schizophrenia. Would anyone like to comment on that?

DR. BIGELOW: We are looking at its effects on the symptoms of schizophrenia, that is what we can measure. We have found in three patients on a blind crossover study that we are seeing some antipsychotic activity, probably by reducing the irritability which, of course, is a common feature shared by manics and schizophrenics. This does not say necessarily that lithium is good for all schizophrenics but it may have some beneficial effects that cannot be found in phenothiazines in alleviating some of the symptoms of schizophrenia.

DR. CANCRO: Well it may vary.

DR. BIGELOW: Yes. This is variable and it has considerable theoretical interest.

Affective features of Schizophrenia

DR. GJESSING: I would comment on Wyatt's Table XVII classifying the action of phenothiazines. The overlap of effect does not surprise me since neither Kraepelin nor Bleuler considered manic depressive or schizophrenic psychoses to be distinct and mutually exclusive. Rather it is a question of how much manic depressive and how much schizophrenia is there in the patient.

DR. ALTSCHULE: Even back in the Renaissance or the Medieval Period affective manifestations were recognized as very common in the early stage of what we today call schizophrenia.

DR. LIPTON: I do not know what happened during the Renaissance, but I think anyone with clinical experience has seen major shifts in affect in early psychosis. The patients are not totally crazy. They know what is happening to them. They sense their social isolation, their defeat and their lost ambitions. They can get significantly clinically depressed. And there is really no reason at all to be surprised by this. We have this curious tendency to isolate symptoms. The schizophrenic patient reported by Freeman who was responsive to folic acid was also mentally retarded (54). There is no reason why you cannot be mentally retarded and schizophrenic. There is no reason why you cannot be schizophrenic and depressed. These things do not necessarily exist in single spectra.

DR. KETY: I would like to comment on this troublesome problem raised by Dick Wyatt, the possibility that we are really dealing with the same, with a single psychosis which is sometimes called schizophrenia and which is sometimes called affective disorder. I could use our data to refute that possibility because in the relatives of the schizophrenics, we find no relationship at all between affective disorder and schizophrenia which I think would settle for me the question as to whether we are dealing with two separate disorders here. Except for one funny thing, we are blind in that study when we make the diagnoses on the relatives. We are blind as to whether they are biological relatives or adoptive relatives, index relatives or control relatives. But we are not blind in one other very important aspect. We know that we are doing a study on schizophrenia. And it is not impossible that we are calling schizophrenia what is just a non-specific psychosis. And the thought had occurred to us that if we thought we were dealing with affective disorder, would we have found the high incidence of affective disorder in the biological relatives. Well we hope to be able to resolve that problem in the affective disorder study because we are going to purposely confound ourselves by mixing in schizophrenic index cases and relatives, so that we will not be looking for only one disorder. I really think that there are striking differences between the people that we call schizophrenic and the people we used to call affective disorder. But I cannot prove it yet.

DR. BLISS: There are family studies which indicate that families with depressive illness have a low incidence of schizophrenia.

DR. KETY: But those studies are not blind. If you know you are studying relatives of depressives, how can you avoid being prejudiced about calling them depressive. As a matter of fact, Steve Matthysse has been going over the old McLean records, studying the pedigrees of patients who came into the McLean with schizophrenia. And the surprising thing is that graduate students who have been reading these say, "These relatives do not all show schizophrenia. They show a variety of psychotic pictures that we would have a hard time calling schizophrenia."

DR. GJESSING: You can make distinctions when you have clear cut cases. But there are cases in which there is a mixture of schizophrenia and affective features.

DR. WOLF: That is right. In every psychiatric hospital there is disagreement about the diagnosis in certain patients. Even the same psychiatrist may change his mind from one month to the next with respect to a diagnosis of schizophrenia versus manic-depressive psychosis.

DR. CANCRO: Yes and even without a change in the patient.

DR. KETY: I have been told that many patients given a diagnosis of schizophrenia at Hopkins by Adolph Meyer, when transferred to Shepherd Pratt were labelled "manic depressive." There is also a difference between English and American psychiatrists with respect to diagnostic preference. Many psychotics who would be considered to be suffering from an affective disorder in America would be called schizophrenic in England. As Dr. Gjessing points out, however, there is always a group of patients with clear cut manifestations in whom all would agree upon a diagnosis of schizophrenia.

"Spontaneous" Remissions

DR. LIPTON: It is overwhelming to encounter someone who has been a back ward patient for 15 or 20 years who then has a lucid week or two. I had an experience like this many years ago when Hudson Hoagland and Charley Huggins got together and decided that a toxic adrenal substance was involved in schizophrenia. I was a resident at the time responsible for the selection of the patients. I found several chronic schizophrenic women who also had carcinoma of the breast, and a couple of men who had carcinoma of the prostate. The carcinoma provided the excuse for doing the adrenalectomy. The surgery made no difference in their psychosis. One of the patients was a severely regressed, back ward schizophrenic who among other things had a greatly enlarged anal sphincter, through which she used to pull feces out of herself. While she was recovering from the adrenalectomy, a depressed lady began to take care of her, put cosmetics on her and the like without, however, benefiting her condition. At the end of some nine months of recuperation from the adrenalectomy she was sent back to the state hospital. A year and a half later I had a call from the state hospital superintendent asking me to come out and see a miracle. The depressed lady had become depressed again and was sent back to the state hospital where she re-established a relationship with the same chronic regressed schizophrenic. At that point, the two of them started what was a sort of homosexual or adolescent romance. The regressed schizophrenic improved to the point where she was reciting Keats and Shelley. When I saw her she said to me "Whatever made you think taking my adrenal out would make a difference?"

As long as the relationship between them lasted, which it did for a couple of weeks, the schizophrenic patient behaved like a normal adolescent. The relationship then broke up. They had a fight, after which the schizophrenic patient regressed again and died that way. But there was a lucid interval of two weeks in a person who had been severely schizophrenic for close to twenty years.

DR. KETY: And that convinced you that schizophrenia was a psychological phenomenon!

DR. LIPTON: No. It convinced me it was a very complex phenomenon, whether biological or psychological it was reversible.

DR. ALTSCHULE: Eldred and Bell did a study about fifteen

years ago and found the way to tell whether a schizophrenic is beginning to improve is not that he begins to relate to other people but that other people begin to relate to him (44). On a closed ward in a fixed population, an occasional patient would begin to improve and you could tell this because the other patients would then begin to talk to him. I think the schizophrenic patient you described was probably beginning to get better when the other woman began to put lipstick on her.

DR. LIPTON: There was no improvement evident to the staff until the second time around.

DR. BIGELOW: I recall your earlier statement saying you really would not study the long term "back ward" patient because this might be an irreversible disease. Does not the patient just discussed contradict your decision?

DR. CANCRO: This field must tolerate contradictions.

CHAPTER IX - Synthesis

Participants: Drs. Altschule, Bigelow, Bliss, Cancro, Cohen, Gjessing, Kety, Lipton, Snyder, Werthessen, Wolf, Wyatt.

DR. CANCRO: It is useful to think of the origin of abnormal behavior in terms of the interaction of independent sources of variance. This is mathematically desirable but certainly not true clinically. The genetic, biochemical, physiologic and psychologic levels are abstractions and not true sources of independent variance. Nevertheless, if we can identify different pathways to the phenotype we will have begun to separate schizophrenia into its component biobehavioral parts. The use of animal models in conjuction with human genealogies would be very helpful in undertaking this particular task.

DR. BIGELOW: I wonder if from the trend of the conversation and taking a lead from the mental retardation story, we should perhaps conclude that any finding, though statistically significant for schizophrenics as compared to reference groups or normals, is unlikely to have anything to do with etiology per se.

DR. CANCRO: You are just being optimistic again.

DR. BIGELOW: Perhaps I am being practical. You are taking a lead from the mental retardation story and every once in a while somebody says, "How many ways are there of being crazy?" or "How many ways are there of being retarded?" Probably only one but there are a host of ways to measure retardation, to establish it beyond reasonable doubt, that it may exist, and does exist in a number of people. But the headway in the syndrome of retardation has come in bits and starts as someone has found a metabolic error and proceeded to work on that theory. So maybe as I tried to show with clinical material, the fact that the very same agent will produce different manifestations in a population presumably homogenous for schizophrenia is further evidence that the population is not homogenous. We should accept the consequences of the lack of homogeneity. This involves probably not looking for a common fault and perhaps saying that when we find a statistically significant common fault in schizophrenia, that may be only pretty far down the chain of causation.

DR. CANCRO: Let me remind you of the difference between etiology and what is secondary to having become ill.

DR. WOLF: I think that Dr. Gjessing was urging this kind of

strategy when he was talking about subclustering.

DR. LIPTON: Subclustering may not teach us about etiology but it can still teach us a good deal about pathogenesis and that is where we are headed.

DR. CANCRO: But before we forget, I would like to comment on a point that Dr. Gjessing made earlier about the 30% recovery. Recently, Zubin (200) reported that his laboratory did a calculation in which they took any psychotic diagnosis such as schizophrenia and assumed it had a validity on the order of .6. They further assumed it had a reliability around the same order of magnitude. They then calculated what would happen if populations of individuals with diagnoses of these orders of reliability and validity were studied. They found that about a third of the cases would get better, a third of the cases would become chronic, and about a third would have histories of recurrent psychotic episodes independent of the diagnosis fixed upon them. It is certainly a remarkable coincidence that this kind of one-third, one-third, one-third distribution keeps coming up over the last hundred years. We have to recognize that part of this may be the syndrome nature of the disorder and the specific problems surrounding the validity and reliability of the diagnosis. Unless there is a significant deviation from 30% one must look cautiously at the figures.

Perception and Performance Tests as Genetic Markers

DR. LIPTON: We do not have a professional psychologist here and I rather wish we did. It strikes me that we limit our possibilities by assuming that if we accept a genetic diathesis the most profitable place to look for a trait marker would be in a protein like an enzyme or in a chemical metabolite. These are useful when we can find them but we infrequently do. Even with Mendelian traits more than 90% of the time we find anatomic, psychologic or behavioral indices rather than chemical ones.

Dr. Gjessing indicated that in a lot of neurological and genetic illnesses a metabolic disorder cannot be found but instead there is disturbed anatomy or disordered function. Some of the psychological findings which stand up, like those of Dave Shakow, on reaction times and in the schizophrenic's limited capacity to shift reaction times, and other findings where estimation of the size of a projected image is measured, must be taken seriously. Schizophrenics with a good prognosis always overestimate the size of a projected image and those with poor prognosis underestimate. Phil Holzman studies eye tracking a swinging pendulum. He finds that schizophrenics show a jagged eye tracking pattern even when they are in remission or on drugs. It therefore appears to be a trait-dependent characteristic rather than a state dependent one. He finds it in the families of schizophrenics at a much higher than chance level. Holzman is now trying to determine whether or not this trait is present in discordant twins. I think he has some preliminary data that in monozygotic twins, the eye tracking pattern is genetically determined and occurs in both twins even when one of them is not schizophrenic. I do not know whether this will stand up. But if it does stand up, it is a real marker!

DR. CANCRO: It has held up in the sense of being replicated. Shagass (164) has done a modification, not yet published, where he took the pendulum and made marks, little numbers, at regular intervals. The subjects were instructed to read the number to themselves as the pendulum passed. When you do this the phenomenon disappears. The task of reading the number appears to increase the subject's attention to the task to the point that the finding disappears. This, in itself, does not wipe out the importance of the result and in fact it may enhance it. The reported phenomenon appears under certain conditions and disappears under others. This makes it a much more interesting model for the schizophrenias than something which is always present. The finding may be actually related to the same parameters Paul Huston et al. (90) used in the reaction time studies done in Worcester. They used the concept of set as the explanation. The results that Holzman reported may also be more related to set than to muscle enzymes such as C P K.

DR. LIPTON: No, but the phenomenon does not disappear when the patient gets well and the phenomenon does not disappear when the patient is on chlorpromazine. It seems to be a trait.

DR. CANCRO: Shagass may still be correct. It may well be a trait for these patients to hypoattend to external tasks unless aroused or motivated to a greater than normal degree.

DR. WOLF: Would you describe the test and indicate the nature of the difference between the schizophrenic and the normal?

DR. LIPTON: Holzman fixed the head and then swung a pendulum in front of the subject asking him to follow it with his eyes. As I recall he had small electrodes on the extraocular muscles and thus obtained an electromyogram of the eye muscles during the to and fro tracking. The normals showed a smooth sinusoidal curve. The schizophrenics showed several jagged curves as if they overshot and then compensated. They did this when acutely or chronically ill and when they were in clinical remission with drugs. They did it on or off drugs and the incidence of abnormal eye tracking was higher in their relatives than it was in the relatives of their normal controls (86).

DR. WOLF: The abnormal pendulum tracking occurs in schizophrenic subjects, is that right?

DR. LIPTON: That is right.

DR. WYATT: But Shagass showed that it was also present in manic depressive patients.

DR. CANCRO: Shagass did find that it was present in the non-schizophrenics as well, but not as markedly as in the schizophrenics. His work tends to place this much more in the attentional sphere. It is not an abnormal finding. It is present in a reasonable per cent, about 10%, of the so-called normal population. But the population of schizophrenics seems to be drawn from that part of the curve. In other words, they are over represented about ten fold.

DR. LIPTON: The point I wanted to make is that it may very well be that the strategy of looking for physiological traits rather than necessarily biochemical traits may still be very useful. In this case, there are lots of hypotheses having to do with the perceptive defects, the "attentional" defects in schizophrenia. Matthysse has worked with the concept of dis-attention, or the inability to detach one's attention from one task and go on to the next job as being characteristic of the schizophrenic. That is -- he just cannot finish the job. This relates to my difficulty with the metabolic or enzymatic findings reported so far. Although schizophrenia is never produced, enzymatic changes can be induced through the use of pharmacologic agents such as inhibitors and precursors that reach at least the order of magnitude found in the brains of schizophrenics. I am not a geneticist and maybe somebody can tell me how often in the well-known genetic disorders we are really able to spot the biochemical defects. Going back to Gregor Mendel is the enzymatic or biochemical defect responsible for wrinkled peas versus smooth peas known? It is a physiological or anatomical marker for peas, and we know it is Mendelian, but what of its biochemistry?

Genetic regulation other than Enzymatic

DR. ALTSCHULE: Biochemical genetics have been developing at an enormous pace. The concentration on the gene controlling an enzyme has been emphasized in total disregard of the manner in which the gene determines form.

DR. LIPTON: You are perfectly correct. For example, my baldness pattern is exactly like that of my father. What enzymes are involved in that? I do not have the slightest idea. There are many enzymes but the form is identical.

DR. ALTSCHULE: It does not have to be enzyme involvement.

DR. COHEN: In sickle-cell anemia hematologists have an idea of how genes are involved in the sickling of red blood cells.

DR. LIPTON: Sure, this is known for sickle-cell anemia and the various hemoglobinopathies. That is a small minority of conditions and in a lot of illnesses we do not know. Therefore we must look at other types of phenomena, physiological dimensions like eye tracking. I have no idea what the enzymology of the eye tracking might be. If we have a trait we have a powerful tool to deal with, a trait which is not state dependent. It is powerful for genetics and it may be that it can give us insight into what some of our integrative difficulties may be. It is a legitimate place to look. We badly need reliable state and trait dependent markers and these need not be biochemical only.

DR. GJESSING: The indication is that the effect is nonspecific for the condition but due to many factors.

DR. LIPTON: I do not know what it means. In the available data, the abnormal pendulum record is present in 25 out of 29 chronic schizophrenics, in 6 of 72 controls, and in 4 of 19 non-psychotic psychiatric patients. Despite the overlap the data are impressive.

DR. WOLF: Since you brought out the question of traits, remembering the work of Sheldon, what can we say today about anthropometric measures which are also a trait, genetically handed down? Does anybody have that on the tip of their tongue?

DR. WERTHESSEN: I can refer you to the work of Crozier and Pincus, published in the early thirties (38). They used two highly inbred strains of rats. Rats exhibit negative geotropism. Placed on a slanting surface, rats go uphill. In this study, young animals were placed on a tiltboard whose angle from the horizontal could be adjusted. Then the angle at which the rat climbed upwards was determined and the interrelation of the two angles determined for each strain. The "dosage" response characteristics of the two strains to the stimulus were different. Pincus crossbred the two strains and studied the progeny. He showed that, like coat color, tropistic responses to physical stimuli are genetically based. The degree of response was a function of the parents' genes in the usual fashion of dominant and recessive traits. Thus the interrelationship of several sensors via coordinating systems within the brain and the expression thereof by predictable movements have been shown to be dependent on the genetic code. I think these are the kind of solid data you needed.

DR. BLISS: There is a large body of information about the genetics of behavior. One can briefly summarize it by saying that if a behavioral trait can be isolated and measured it can usually be accentuated or reduced through various inbreeding experiments.

DR. WERTHESSEN: This is the whole argument in breeding dogs

for hunting birds.

DR. BLISS: Exactly.

DR. LIPTON: It is a complex type of behavioral trait about which we know little because it is not in a single discipline. Interdisciplinary programs ought to concentrate and pay attention to such problems because they may yield rich rewards. We focus on our kind because we have the power to alter the enzymology pharmacologically but it is more difficult to alter behavior chronically. Though we cannot even be sure of that. I do not know what amphetamine or chlorpromazine will do to the behavior of specific strains of rats or mice. It may be worth looking into.

Physiologic Approaches

DR. CANCRO: Let me try to bring the discussion back to physiology. I would like to emphasize again that in the physiology you are looking at a different level of abstraction regarding the way the nervous system functions. We need an analogy and the one of a wiring diagram comes to mind. I hope none of us take that too seriously. Unfortunately with this analogy you may get the picture of the nervous system as a fixed system which if it is wired in a particular way there is nothing you can do about it. It is better to think in terms of preferential wiring patterns which can be abnormal when altered. The importance of the physiologic or psychophysiologic measures is that they can serve as an interface between the clinical observations and biochemical measures. Philip Holzman is continuing to work on this despite the fact that Shagass' finding has caused some people to reject the rotary pursuit research. The investigation remains significant and should be pursued. In passing, I will mention a couple of things that have come out of our own laboratory recently which need to be replicated and need to be studied more carefully.

In comparing a group of schizophrenic patients, hospitalized depressive patients, and normal controls, we found that during free scan, visual and mental tasks, the normal controls showed a certain pattern in terms of fixation, blinking and eye movements (23). It varied as a function of the amount of information reaching the retina. It was very reasonable in the sense that when the subjects performed visual tasks, the amount of information reaching the retina increased in terms of the proportion of fixation time and when the subjects were doing mental tasks the amount or proportion of fixation time was reduced. The hospitalized depressives followed the identical pattern and showed no significant difference. The schizophrenics showed a significant difference of about one standard deviation in all conditions. They were out of phase in all conditions. They were less interested in the visual environment, even though they cooperated with the investigators in a reasonable way. In a sense they attended less to the visual environment and were more absorbed in their thoughts. During rest or free scan their basal level was at exactly the same point as that of the normals and depressives during mental tasks. We are also looking at the degree of determination of eye movements in these tasks. As you would expect, eye movements were not random in normals. They are also not random in depressives according to the Markoff method of analysis which we used (24). They are interestingly even more determined in schizophrenics. This we had not expected. We went back and looked at the data more carefully and especially the scoring technique. The scoring technique involves giving a particular score if, for example, a person looks to the left and then he follows that with another leftward movement, a low probability event. It turned out that schizophrenics tend to perseverate in their eye movements. When looking to the left they continue to look to the left, regardless of the task they are doing. If they started to look up, they tend to continue to look up. This particular finding fits the psychology literature which reports that schizophrenics tend to perseverate in their responses as a function of prior stimuli. This is our finding here too, at least in the first part of the sample. They seem to perseverate prior to and during the task and do this independently of the task requirement. In terms of "determination" the movements are highly determined but the term is misleading. We can see the behavior as being an inefficient manner of processing visual information. To hypothesize for a moment, I suspect this way of attending to the visual environment is more of a trait than it is related to the state. We do not know yet and will have to do longitudinal studies. I suspect we are finding an attentional pattern which is more common in people who develop a schizophrenic illness. These kinds of strategy, these kinds of approaches make very good sense but again they only make genuine sense when they are integrated into an interdisciplinary program of research. This program ideally would proceed with longitudinal studies using the patient as his own control particularly when he goes into remission, when he relapses, etc.

I would say that these are high personal priorities for me. I would pursue longitudinal research taking a problem oriented interdisciplinary approach and using clinical state, physiologic findings, or combinations of these as markers to identify moments in time when the team should in a sense descend on the patient.

DR. KETY: No one would argue that one should be monolithic in one's approach. One should not neglect the psychologic, the social or the clinical area. There is, however, something to be said for the enthusiasm now surrounding the biochemical approach. Biochemists are exploring new fields that have not been explored before. The great advantage of the biochemical approach to schizophrenia is that these biochemical mechanisms are brand new. There are new tools which have not been used before. The psychologists and the clinicians have been looking at schizophrenia for a long time; that does not deny the possibility that somebody may come up with something new in that area, but in terms of simple probability one is likely to come up with more new things in an area only recently opened up, than in an area which has been explored over and over again. I would not want my comments to mean a derogation of any other kind of research. The important thing is for each investigator to pick the area in which he feels he can most likely contribute in terms of his confidence in his strategy.

Biochemical Approaches

DR. LIPTON: I certainly did not want to de-emphasize biochemistry. Work on immune assays for the hypothalamic hormones, may furnish a new light on many subjects. These methods have only become available recently and there are many exciting directions in which one can go. In one sense the transmitter area is getting to be almost old hat. It has been popular for ten years or more. And while new micro-methods are coming up in enzymology or measurements are getting down to nanogram quantities, or even fractions of nanograms, I do not think we ought to limit ourselves to just the traditional transmitter studies. Study of the newer transmitters and of receptors should prove profitable.

DR. WOLF: One should be able to sharpen the biochemical approach through this interdisciplinary approach and multinational cooperation. If one could have a relatively homogeneous group to study like the Murphree dogs for example, one might be able to perceive more acutely the information obtainable from the biochemical study.

By way of a general philosophical remark, we probably do not exploit that kind of opportunity as much as we should in the way that Dr. Kety did in the genetics study.

DR. LIPTON: Another interesting problem has to do with vitamin dependency diseases. In 1970 there were 12 of them. In 1974 there were 25. I do not know how many there will be by 1980. So much for vitamins. In theory there is no reason why there should not be mineral dependency diseases. The point I want to make has to do with technology and I think it is an important one. Vitamin dependency diseases are rare. They appear about one in 40,000 births. But that means the heterozygote occurs in probably 1 in 200. If you have 25 illnesses in which the heterozygote is appearing one in 200, that means one in eight births is likely to be a heterozygote for one of these conditions. The technology for studying the heterozygote is just beginning. It involves tissue culture and the capacity to study partial enzymes; it is very difficult.

I do not have the slightest idea of what will turn out to be the nutritional requirements of the heterozygote. This might be relevant to the issue of why there is a 30 fold variation in the levels of imipramine. It may be another area well worth looking at because at least one vitamin deficiency illness manifested itself as schizophrenia. In many of the vitamin deficiency diseases there are convulsions, mental retardation and many neurological symptoms. Heterozygotes may have susceptibilities that we do not know about. They may not be terribly uncommon.

DR. WOLF: You are saying that the pattern of disorder may be entirely different.

DR. LIPTON: Yes.

DR. WOLF: I think that is an important point.

DR. ALTSCHULE: One thing about working in a library is that you develop chronic, intermittent déjà vu. There is no need to particularize. If you have not worked out something with an old technique, you had better get a new one as new ones are more acceptable. However, you tend to form the same patterns. The fact that patients with schizophrenia show so many disturbances in what we consider psychic function, makes one naturally turn his attention to the brain. Yesterday someone pointed out that most of the symptoms in PKU are traceable to a liver disorder and therefore it might be appropriate and worthwhile to study schizophrenia in terms of possible disorders of transmitters. It is entirely plausible and useful to pursue other possibilities. One of the possibilities now being taught to internists is that something about mechanisms may be learned from the reaction to medication. I do not mean to imply that we should conclude that pneumonia is caused by a genetically determined lack of penicillin in the body or that chronic constipation patients should be considered to have hypocascarism. Nevertheless there are clues. Cancer chemotherapy is very largely dependent upon the concept of competitive inhibition. Has anybody considered the possibility of competitive inhibition in terms of schizophrenia?

Metabolic Features

One of the striking things about schizophrenia is that while the patient responds favorably to phenothiazines it is not that the drugs are merely sedative or anti-anxiety drugs; they have a specific effect on psychotic manifestation. If you look through the natural history of compounds which resemble phenothiazine and are normally present in mammals, you find one.

We know that the amino acid tryptophan changes to the amine form as serotonin and melatonin. Tryptophan is metabolized in the normal fashion to anthranilic acid, some of it being excreted as such and some as nicotinic acid. Work done about 15 years ago showed that at least in the rat liver, some anthranilic acid, by the conjugation of two molecules, forms one or more of a class of three-ringed molecules. These resemble phenothiazines with respect to the position of the ring nitrogen but instead have ring oxygen in place of the phenothiazine sulphur. The addition of peptides to such a molecule leads to the formation of a highly toxic actinomycin moiety. The anthranilic acid conjugates are also found in grasshopper's eyes. I wonder why this work has not received as much attention as I think it deserves (176). Here you have a natural compound made in the liver which in the related actinomycin form is highly poisonous and could very well participate in competitive inhibition with phenothiazines.

DR. BIGELOW: But I thought they were in lobster eyes.

DR. ALTSCHULE: Either one will do!

DR. WOLF: Mark has a very strong sense of priorities.

DR. ALTSCHULE: Has anybody here thought about this before?

DR. WYATT: In a different way. Let me show a different structure. If you take this tetrahydromethylfolate, tryptamine or methyltryptamine, after incubation you end up with a structure that looks like Fig. 27, and resembles if you wanted to, reserpine probably more than it does chlorpromazine but does closely resemble chlorpromazine as well. You can play with molecules.

DR. ALTSCHULE: Yes, also that formula resembles the cyclized form of melatonin.

DR. SNYDER: There is one big difference between carboline and phenothiazines. A carboline ring system is planar. With phenothiazine the rings are tilted toward each other, which is crucial for antischizophrenic actions.

DR. CANCRO: Since we now have convincing evidence of the genetic transmission of schizophrenia, increasingly the focus must shift from disciplinary arguments as to nature-nurture to identifying mechanisms, to identifying traits, and to getting at the particular issues which may be relevant to the disorder. Animal models offer promising possibilities. Matthysse in 1974 wrote an interesting paper on the role of animal models in schizophrenia research (119). It is a very sophisticated approach where he speaks of the need to define mechanisms rather than find isomorphic representations of schizophrenia in animals. It is not necessary to have a schizophrenic rat or a schizophrenic monkey;

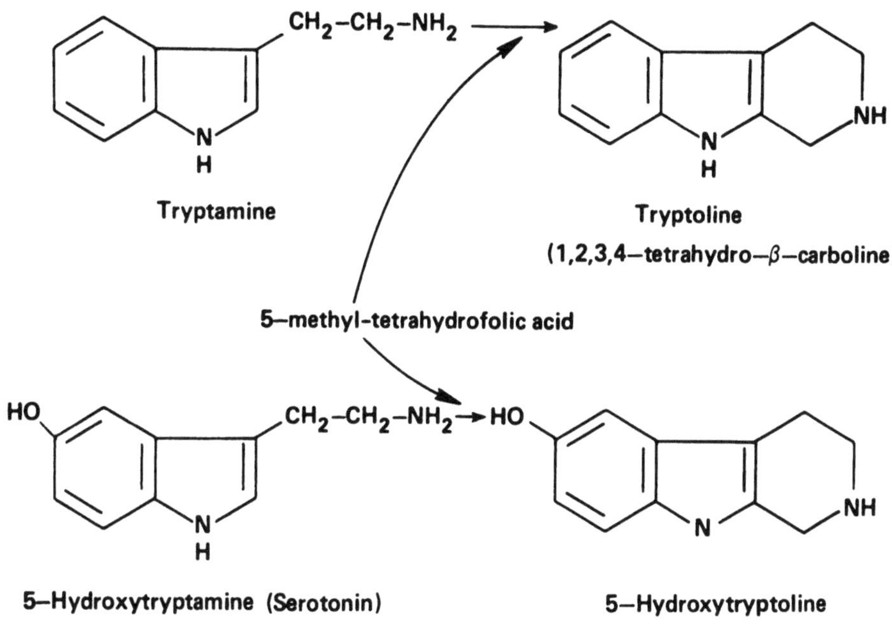
CH2-CH2-NH2
N
H
Tryptamine
NH
N
H
Tryptoline
(1,2,3,4-tetrahydro-β-carboline)
5-methyl-tetrahydrofolic acid
HO
CH2-CH2-NH2
HO
N
H
NH
N
5-Hydroxytryptamine (Serotonin)
5-Hydroxytryptoline

FIG. 27

all you need is some mechanism that is homologous and that can be studied genetically and in other ways.

Psychophysiological Approaches

The study of patients through psychophysiologic approaches also offers some very important and powerful strategies that can greatly strengthen our understanding the behavior at one end of the continuum and the biochemical phenomenon if you will, at the other end of the continuum. The use of psychophysiological approaches in identifying parameters that are affected or that, at the least, seem to differ can be a very useful strategy. For example, if you look at schizophrenia as a disorder in the regulation of the understanding and processing of information, then you can seek to identify times when this person is processing information in a deviant way even if this is only a statistically deviant fashion. Having identified such a time it would then be extremely useful and appropriate to make the biochemical determination. In this way you are not faced with the problem of not knowing where you are functionally when the samples are obtained. The point has been emphasized again and again indirectly, and needs to be made explicitly in pointing out the variability in the syndrome.

We have talked about thought disorder but unlike organic dementia, thought disorder in schizophrenia is intermittent. A patient given a series of proverbs abstracts beautifully on the first two; on the third one he gets a bit bizarre and on the fourth one totally confused the observer. On the fifth proverb he abstracts again beautifully. Thought disorder is a rather curious phenomenon. Virtually all of the phenomena that we see in schizophrenia, wax and wane, being more evident at some times than others. Not only does the patient fluctuate, but even nonschizophrenics fluctuate from moment to moment. Schizophrenics seem to fluctuate more dramatically than normals. It is important that we move away from some of the psychological-organic controversy because it tends to obscure our thinking. A good example of this is what happened with some of the reaction-time studies. When it became clear that the patient could be punished into a normal reaction time range, the finding was seen as motivational in origin and no longer interesting. The delay in reaction time was attributed to negativism or disinterest. It was seen as part of the psychology of the disorder. Yet it may be a very productive way of approaching this disorder because the illness has this characteristic of fluctuating and lack of constancy. I do not think a study of reaction time is useful in isolation but when it is integrated with these other approaches, it can be very helpful.

There is a need for a multiplicity of approaches. Perhaps the best would be an interdisciplinary approach oriented toward a particular problem where the efforts can enhance and interact with each other. I think this has to be distinguished from a crash program. Dr. Altschule and I were discussing crash programs briefly yesterday and without being concrete, they seem to be characterized by a remarkable ability to crash. Crash programs cannot develop successfully in Washington on something like cancer, schizophrenia or poverty. They just do not lend themselves to the crash program approach.

DR. ALTSCHULE: Maybe a crash program to study crash programs?

DR. CANCRO: In all seriousness, I think we must avoid the bandwagon crash program approach which states: "We must stamp out schizophrenia this summer and then do something else in the fall." Yet if we do not maintain a focus on the problem there is another danger. The patient and his problems may become lost.

It is, I think, vital that we not lose sight of the patient in schizophrenia research. This is a truism. Unfortunately, it is also a truism that as one gets involved in one's own particular highly specialized area of research, the mechanisms become interesting, the problems become interesting and one finds oneself tending to move more and more into a tunnel of following up on the last set of leads. Laboratories develop a certain momentum of their own, a certain life of their own. The laboratories have to keep functioning and operating. The equipment has to be maintained and the people have to be fed. Granting agencies have their own peculiar ideas of where the work should go so that it is very easy in schizophrenia research to begin to be productive and then wander away from the patient and get involved in the study of basic mechanisms. These of course are important. I think that part of what can help to maintain this patient focus is the involvement of genuine, real, clinical people in schizophrenia research. They can help to keep clinical questions and clinical realities in focus in our laboratories so that we do not wander away too far. It is also important to recognize that the clinical state of a patient can be used in much the same way as a radioactive isotope is used. That is, the patient is followed over a time; biochemical and psychophysiological differences are sought when he is acutely hallucinating, delusional, in remission, or symptom free. What are the trait dependent variables; what are the state dependent variables? The clinical picture can be very useful to indicate this type of marker and has not been used in that way, partially because of a paucity of longitudinal research in schizophrenia.

The last remark I would like to make is that it is very important to make a distinction in our thinking between pathways and mechanisms which may contribute to the etiology of a disorder and mechanisms which may be a consequence of the disorder. Cross-sectional studies cannot disentangle these differences, which point brings us back to the need for good longitudinal studies.

DR. WOLF: Perhaps it would be fair to say that we have learned something of the distribution of schizophrenic manifestations, something of the intermediary neurochemical mechanisms and something of genetic influences on etiology. A parallel to diabetes has been suggested. The chemical metabolic aberration responsible for the symptoms and findings may be induced by a variety of stimuli including drugs and extreme situational stresses. On the other hand, at least in one cluster of manifestations called schizophrenia, there is a strong genetic proclivity toward the chemical metabolic aberrations of the schizophrenic process. Combinations of these several factors may work through a final common neurochemical pathway that we are just beginning to understand and to some extent can already manipulate.

BIBLIOGRAPHY

1. Achte, K.A., Hillbom, E., and Aalberg, V.: Psychoses following war brain injuries. Acta Psychiat. Scan., 45: 1-18, 1969.

2. Aghajanian, G.D. and Bunney, B.S. Central dopaminergic neurons: Neurophysiologic identification and responses to drugs. In Frontiers in Catecholamine Research, E. Usdin and S. H. Snyder, Eds. Pergamon Press, Oxford, 643-648, 1974.

3. Altschule, M.D. (Ed.): Some effects of aqueous pineal extract administration on the symptoms of schizophrenia. Frontiers of Pineal Physiology. Chapt. 9. M.I.T. Press, Cambridge, Mass. 1975.

4. Altschule, M.D.: The Development of Traditional Psychopathology -- A Source Book. In preparation.

5. Ashkenazi, R. and Haber, B.: Inhibition of synaptosomal biogenic amine uptake by sympathomimetic amines. Proc. Am. Soc. Neurosc. 29: 123, 1974.

6. Ashkenazi, R. and Haber, B.: The metabolism of phenylethylamine and O-methylated derivatives by monoamine oxidase. Experientia, 31: 335, 1975.

7. Ashkenazi, R. and Haber, B.: Mode of action of a sympathomimetic amine, paramethoxy-phenylethylamine (PMPEA) in the mouse CNS: release of biogenic amines. Neuroscience Letters, 1: 163-168, 1975.

8. Axelrod, J.: The enzymatic N-methylation of serotonin and other amines. J. Pharmacol. Exp. Ther. 138: 28-33, 1962.

9. Ball, B.: Lecons sur les maladies mentales. pp. 238-239, Asselin et Ce, Paris, 1880.

10. Belmaker, R., Beckman, H., Goodwin, F., Murphy, D., Pollin, W., Buchsbaum, M., Wyatt, R.J., Ciaranello, R., and Lamprecht, F.: Relationships between platelet and plasma monoamine oxidase, plasma dopamine-beta-hydroxylase and urinary 3-methoxy-4-hydroxy phenylglycol. Life Science 16: 273-280, 1975.

11. Belmaker, R., Murphy, D., Wyatt, R.J., and Lorieux, D.L.: Human platelet monoamine oxidase changes during the menstrual cycle. Arch. Gen. Psychiat. 31: 537-541, 1974.

12. Boklage, C.E.: Schizophrenia, brain asymmetry development, and twinning: Cellular relationship with etiological and possibly prognostic implications. In press.

13. Böök, J.A.: A genetic and neuropsychiatric investigation of a North Swedish population with special regard to schizophrenia and mental deficiency. Acta Genet. (Basel) 4: 1. 1953.

14. Borenstein, P., Dabbah, M., and Metzger, J.: L'encephalographie fractionnéé dans les syndromes schizophreniques. Ann. Med. Psychol., 2: 385-426, 1957.

15. Bowers, M.D.: Pathogenesis of acute schizophrenic psychosis. Arch. Gen. Psychiat., 19: 348-355, 1968.

16. Bratfos, O., and Sagedal, E.: Luftencefalografiske undersokelser hos pasienter innlagt i sinnssykehus. Nord. Med., 51: 1606-1609, 1960.

17. Burrows, G.M.: An inquiry into certain errors relative to insanity and their consequences: physical, moral and civil. Thomas and George Underwood, London, 164-165, 484-489, 1820.

18. Burton, R.: Anatomy of Melancholy, 1610. Edited by Floyd Dell and Paul Jordan-Smith, Tudor Publishing Co., New York, 1927.

19. Calhoun, J.B.: Population Density and Social Pathology. Scientif. Am., 206, 2: 139-148, 1962.

20. Campion, E., and Tucker, G.: A note on twin studies, schizophrenia and neurological impairment. Arch. Gen. Psychiat., 29: 460-464, 1973.

21. Canavan, M. and Clark, R.: The mental health of 463 children from dementia praecox stock. Mental Hygiene, 7: 137-148, 1923.

22. Cancro, R.: Prospective Prediction of Hospital Stay in Schizophrenia. Arch. Gen. Psychiat., 20: 541-546, 1969.

23. Cancro, R.: Visual Attention in Schizophrenia (Abs.) Scient. Proc. 128th Ann. Mtg. A.P.A. Washington, A.P.A. 118-119, 1975.

24. Cancro, R., and Glazer, W.: A Markoff Analysis of Eye Movements in Schizophrenics, Hospitalized Depressives and Normal Controls. Unpublished manuscript, 1975.

25. Carenzi, A., Gillin, J.C., Guidotti, A., Schwartz, M., Trabucchi, M. and Wyatt, R.J.: Dopamine-sensitive adenyl cyclase in human caudate nucleus. Arch. Gen. Psychiat. 32: 1056-1059, 1975.

26. Carlsson, A., and Lindqvist, M.: Effect of chlorpromazine or haloperidol on formation of 3-methoxytyramine and normetanephrine in mouse brain. Acta Pharmacol. Toxicol. 20: 140-144, 1963.

27. Carpenter, W.T. Jr., Murphy, D.L. and Wyatt, R.J.: Platelet monoamine oxidase in acute schizophrenia. Am. J. Psychiat. 132: 438-441, 1975.

28. Carroll, L.: Symbolic Logic. Dover, New York, p. 2, 1958.

29. Chapman, J.: The early symptoms of schizophrenia. Brit. J. Psychiat., 112: 225-251, 1966.

30. Charpentier, P.M.A.: Les démences précoces. Compte Rendu du Congrès Annuel de Médecin Mentale, Première session. G. Masson, Paris, pp. 247-268, 1890.

31. Codoret, R.J.: Toward a definition of the schizoid state: Evidence from studies of twins and their families. Brit. J. Psychiat., 122: 679-685, 1973.

32. Colon, E.J.: Quantitative cytoarchitectonics of the human cerebral cortex in schizophrenic dementia. Acta Neuropathol. (Berlin) 20: 1-10, 1972.

33. Conolly, J.: An inquiry concerning the indications of insanity. John Taylor, London, pp. 125-127, 1830.

34. Corson, S.A.: The violent society and its relation to psychopathology in children and adults. In Society, Stress and Disease, Vol. 2 (Lennart Levi, Ed.) Oxford University Press, 1975.

35. Cravioto, J., Hambraeus, L. and Vahlquist, B. (Eds).: Early Malnutrition and Mental Development Symposia of the Swedish Nutrition Foundation XII. Almquist & Wiksell, Uppsala, Sweden, 1974.

36. Creese, I., Burt, D.R., and Snyder, S.H.: Dopamine receptor binding: Differentiation of agonist and antagonist states with ^{3}H dopamine and ^{3}H haloperidol. Life Sciences, In Press.

37. Crichton, A.: An inquiry into the nature and origin of mental derangement. Cadell, T., and Davies, W., London, p. 344, 1798.

38. Crozier, W.J., and Pincus, G.G.: Analysis of the Geotropic Orientation of Young Rats. Studies 5 and 6. J. Gen. Physiol. 15: 421-462, 1932.

39. Davis, P.A.: Evaluation of the electroencephalograms of schizophrenic patients. Am. J. Psychiat., 96: 851-860, 1940.

40. Denckla, M.B.: Syndromes in Learning Disabilities: The case for "splitting" versus "lumping." J. Learning Disabilities, 5: 26-32, 1972.

41. Dickson, J.T.: The Science and Practice of Medicine in relation to mind: The pathology of nerve centers and the jurisprudence of insanity. D. Appleton and Co., New York, pp. 290-292, 1874.

42. Dubos, R.: Mirage of Health. Harper and Bros., New York, 1959.

43. Dufour, J.M.: Essai sur les opérations de l'entendement humaine et sur les maladies. Merlin, Amsterdam, p. 357, 1770.

44. Eldred, S.H., Bell, N.W., Sherman, L.J., and Langabaugh, R.H.: Classification and analysis of interation patterns on a ward for chronic schizophrenics. Research Publications, Assoc. for Res. in Nerv. and Ment. Dis., D.McK. Rioch and E.A. Weinstein, Eds., Baltimore, 1964.

45. Erlenmeyer-Kimling, L.: Progress report (1973) In: Children at risk: The search for the antecedents of schizophrenia. Garmezy, N., Ed. Schizophrenia Bulletin, 8: 14-90, 1974.

46. Erlenmyer-Kimling, L., Rainer, J.D., and Kallmann, F.J.: Current Reproductive Trends in Schizophrenia. In: Psychopathology of Schizophrenia. Hoch, P.H. and Zubin, J. (Eds.), Grune and Stratton, New York, 1966.

47. Esquirol, E.: Mental Madness: A treatise on insanity. Lea and Blanchard, Philadelphia, pp. 417-418, 1845.

48. Feinberg, A.P. and Snyder, S.H.: Phenothiazine drugs: Structure activity relationships explained by a conformation that mimics dopamine. Proc. Natl. Acad. Sc. USA 72: 1899-1903, 1975.

49. Fischer, M.: Psychoses in the offspring of schizophrenic monozygotic twins and their normal co-twins. Brit. J. Psychiat., 118:43-52, 1971.

50. Fish, B. and Alpert, M.S.: Abnormal states of consciousness and muscle tone in infants born to schizophrenic mothers. Am. J. Psychiat., 119:439-445, 1962.

51. Fisher-Perrudon, C.: Insomnie totale pendant pleusieurs mois et métabolism de la serotonine. (Thesis) J.Tixier & Fils, Lyon, 1973.

52. Frankova, S.: Malnutrition and stimulation in animals. In: Early Malnutrition and Mental Development. Symposia of the Swedish Nutrition Foundation XII (J. Carvioto, L. Hambraeus, B. Vahlquist, Eds.) Almquist & Wiksell, Uppsala, pp. 202-210, 1974.

53. Freedman, B., and Chapman, L.J.: Early subjective experience in schizophrenic episodes. J. Abn. Psychol., 82: 46-54, 1973.

54. Freeman, J.M., Finkelstein, J.D., and Mudd, H.S.: Folate-responsive homocystinuria and "schizophrenia." New Eng. J. Med. 292: 491-496, 1975.

55. Fremming, K.H.: The expectancy of mental disorders. (Thesis) Copenhagen, 1947., and The Expectation of Mental Infirmity in a Sample of the Danish Population, London, 1951.

56. Freud, S.: Psychoanalytic notes upon an autobiographical account of a case of paranoia (dementia paranoides) (1911) Collected Papers, Vol. 3, Hogarth Press, London, 1949.

57. Galton, F.: Hereditary Genius. MacMillan and Co., London, 1892.

58. Gantt, W.H.: Principles of nervous breakdown - schizokinesis and autokinesis. In: Comparative Conditioned Neuroses. Ann. N.Y. Acad. Sci., 56: 147-168, 1953.

59. Garelis, E., Gillin, J.C., Wyatt, R.J., and Neff, N.: Elevated blood serotonin concentrations in chronic, unmedicated schizophrenics: A preliminary study. Am. J. Psychiat. 132: 184-186, 1975.

60. Garmezy, N.: Children at risk: the search for the antecedents of schizophrenia. Part 2. Schizophrenia Bulletin, 9: 55-125, 1974.

61. Gauthier, G.C.: De la démence précoce chez les jeunes alienes hereditaires. (Thesis) University of Paris, No:376, 1883.

62. Georget, M.: De La Folie. Considérations sur cette Maladie. Chez Crevot, Paris, p. 119, 1820.

63. Giannitrapani, D. and Kayton, L.: Schizophrenia and EEG spectral analyses. EEG and Clin. Neurophysiol. 36: 377-386, 1974.

64. Gillin, J.C., and Wyatt, R.J.: The psychedelic model of schizophrenia: The case of N,N-dimethyltryptamine (DMT). In Press.

65. Ginsburg, B.E.: A Dog Model of Hyperkinesis. Unpublished manuscript, 1975.

66. Goldfarb, W.: The subclassification of psychotic children: Application to a study of longitudinal change. In: The Transmission of Schizophrenia (Rosenthal, D. and Kety, S. Eds.), p. 333, Pergamon Press, Oxford, 1968.

67. Goldfarb, W.: Childhood Schizophrenia. Harvard Press, Cambridge, Mass., 1961.

68. Goldstein, L., Sugerman, A.A., Stolberg, H., Murphree, H.B., and Pfeiffer, C.C.: Electrocerebral activity in schizophrenics and non-psychotic subjects: quantitative EEG amplitude analysis. EEG and Clin. Neurophysiol., 19: 350-361, 1965.

69. Gooch, R.: The most important diseases peculiar to women: With other papers. New Sydenham Society, London, pp. 58-59, 1859.

70. Gottesman, I.I. and Shields, J.: Schizophrenia in twins: Sixteen years' consecutive admissions to a psychiatric clinic. Brit. J. Psychiat. 112: 809, 1966.

71. Griesinger, W.: Mental Pathology and Therapeutics. (Translated by C. Lockhart Robertson and James Rutherford). New Sydenham Society, London, p. 83, 1967.

72. Guislain, J.: Traité sur les phrénopathies, ou doctrine nouvelle des maladies mentales. Etablissement Encyclographique, Brussels, pp. 258-260, 1833.

73. Guislain, J.: Traité sur les phrénopathies, ou doctrine nouvelle des maladies mentales. Etablissement Encyclographique, Brussels, p. 325, 1833.

74. Hallgren, B., and Sjogren, T.A.: A clinical and genetico-statistical study of schizophrenia and low grade mental deficiency in a large Swedish rural population. Acta Psychiat. Neurol. Scand. Suppl., 140: 35, 1959.

75. Hartley, D.: Observations on Man. Johnson, London, 1801.

76. Haslam, J.: Observations on Madness and Melancholy. J. Callon, London, pp. 64-67, 1809.

77. Haslam, J.: Illustrations of Madness: Exhibiting a singular case of insanity, and a no less remarkable difference in medical opinion: Developing the nature of assailment, and the manner of working events; with a description of the tortures experienced by bomb-bursting, lobster-cracking, and lengthening the brain. G. Hayden, London. 1810.

78. Haug, J.O.: Pneumoencephalographic studies in mental disease. Acta Psychiat. Scand., 37 (Suppl. 165) 11-104, 1962.

79. Hecker, E.: Die Hebephrenie. Virchow's Archiv. fur Pathologische Anatomie, 52: 394-429, 1871.

80. Helgason, T.: Epidemiology of mental disorders in Iceland. A psychiatric and demographic investigation of 5395 Icelanders. (Thesis). Munksgaard, International Booksellers and Publishers Ltd., Copenhagen, 1964.

81. Hertzig, M.A., and Birch, H.G.: Neurologic organization in psychiatrically disturbed adolescent girls. Arch. Gen. Psych. 15: 590-599, 1966.

82. Hertzig, M.A., and Birch, H.G.: Neurologic organization in psychiatrically disturbed adolescent girls. Arch. Gen. Psych. 19: 528-537, 1968.

83. Heston, L.L.: Psychiatric Disorders in Foster Home Reared Children of Schizophrenic Mothers. Brit. J. Psych. 112: 819-825, 1966.

84. Heston, L.L.: Genes and Schizophrenia. In Biological Psych. Mendels, J. (Ed.) John Wiley & Sons, New York, 1973.

85. Hollister, L.E.: Chemical Psychoses-LSD and related Drugs. Charles Thomas, Springfield, Ill., 1968.

86. Holzman, P.H., Proctor, L.R., Levy, D.L., Yasillo, N.J., Meltzer, Y.H., and Hurt, S.W.: Eye tracking dysfunctions in schizophrenic patients and their relatives. Arch. Gen. Psych. 31: 143-151, 1974.

87. Honigfeld, G. and Klett, C.J.: The Nurses' Observation Scale for Inpatient Evaluation. J. Clin. Psychol. 21: 65-71, 1965.

88. Huber, G.: Pneumoencephalographische und psychopathologische bilder bei endogenena psychosen. Springer, Berlin, 1957.

89. Hurst, L.A.: Electroencephalographic support for a genetically oriented organic concept of schizophrenia. J. Nerv. & Ment. Dis., 115: 90-120, 1952.

90. Huston, P.S., Shakow, D., and Riggs, L.A.: Studies of motor function in schizophrenia: II. Reaction Time. J. Gen. Psychol., 16: 39-82, 1937.

91. Itil, T.M.: Progress report in Children at risk: the search for the antecedents of schizophrenia. Part 2. Schizophrenia Bulletin, 9, 1974.

92. Itil, T.M., Saletu, B. and Davis, S.: EEG findings in chronic schizophrenics based on digital computer period analyses and analog power spectra. Biol. Psychiat. 5: 1-13, 1972.

93. Jacobi, W., and Winkler, H.: Encephalographische studien an chronisch schizophrenen. Arch. Psychiat. Nervenkr., 81: 299-332, 1927.

94. Jacquelin-Dubuisson, J.R.: Des Vésanies, ou Maladies Mentales. Chez Pauteur, Paris, pp. 249-251, 1816.

95. Janowsky, D.S., El-Yousef, M.K., Davis, J.M., and Sekerke, H.J.: Provocation of schizophrenic symptoms by intravenous administration of methylphenidate. Arch. Gen. Psych. 28: 185-191, 1973.

96. Johnson, R.T.: Experimental rabies. Studies of cellular vulnerability and pathogenesis using fluorescent antibody staining. J. Neuropath. Exptl. Neurol. 24: 662-674, 1965.

97. Johnson, R.T.: Mumps virus encephalitis in the hamster. Studies of the inflammatory response and non-cytopathic infection of neurons. J. Neuropathol. Exptl. Neurol. 27: 80-95, 1968.

98. Kahlbaum, K.: Die Katatonie oder das Spannungsirresein. Eine klinische Form psychischer Krankheit. August Hirschwald, Berlin, 1874.

99. Kahlbaum, K.: Ueber jungendliche Nerven - und Gemuthskranke und ihre podagogische Behandlung in der Heilanstallt. Allgemeine Zeitschrift fur Psychiatrie 40: 868-873, 1884.

100. Kallman, F.J.: The Genetics of Schizophrenia. J. J. Augustin, New York, 1938.

101. Kaplan, J., Stillman, R., Gillin, J.C., Walker, R., Mandel, L., and Wyatt, R.J.: Absence of tolerance to DMT in normal experienced human subjects. In Press.

102. Karlsson, J.L.: Genetic Association of giftedness and creativity with Schizophrenia. Hereditas, 66: 177-182, 1970.

103. Kebabian, J.W., Petzold, G.L., and Greengard, P.: Dopamine-sensitive adenylate cyclase in caudate nucleus of rat brain and its similarity to the "dopamine receptor." Proc. Natl. Acad. Sci. USA 69: 2145-2149, 1972.

104. Kennard, K.: Value of equivocal signs in neurological diagnosis. Neurology, 10: 753-764, 1960.

105. Kety, S.S., Rosenthal, D., Wender, P.H. and Schulsinger, F.: The types and prevalence of mental illness in the biological and adoptive families of adopted schizophrenics. In: The Transmission of Schizophrenia, D. Rosenthal and S. S. Kety (Eds)., Pergamon Press, Oxford, pp. 345-362, 1968.

106. Kety, S.S., Rosenthal, D., Wender, P.H., Schulsinger, F., and Jacobsen, B.: Mental illness in the biological and adoptive families of adopted individuals who have become schizophrenic: A preliminary report based upon psychiatric interviews. In Genetic Research in Psychiatry, R. Fieve, D. Rosenthal and H. Brill (Eds)., The Johns Hopkins University Press, Baltimore and London, pp. 147-165, 1975.

107. Kovacic, B., and Domino, E.F.: Tolerance to behavioral effects of dimethyltryptamine (DMT) in the rat. Fed. Proc., 33: 549, 1974.

108. Kraepelin, E.: Lehrbuch der Psychiatrie. VI. Edited, translated abstracted and adapted as Clinical Psychiatry by A. R. Defendorff. The MacMillan Company, New York, pp. 152-160, 1904.

109. Kraepelin, E.: Dementia Praecox and Paraphrenia. Translated by R. M. Barclay. E & S Livingston, Edinburgh, 1919.

110. Krafft-Ebing, R. von: Textbook of Insanity. F. A. Davis, Philadelphia, 1904.

111. Laitenen, L.: Stereotaxic lesions in the knee of the corpus callosum in the treatment of emotional disorders. Lancet, 1: 472-475, 1972.

112. Larsen, V.: Physical characteristics of disturbed adolescents. Arch. Gen. Psychiat., 10: 55-54, 1964.

113. Lasèque, C.: Du delire de persecutions. Arch. Gen. de Medecine 40 Serie 28: 129-150, 1852.

114. Lemke, R.: Untersuchungen uber die soziale prognose der schizophrenie unter besonderer beruck sichtigung des encephalographischen befundes. Arch. Psychiat. Nervenk., 104: 89-136, 1935.

115. Lester, B.K. and Edwards, R.J.: EEG fast activity in schizophrenic and control subjects. Int. J. Neuropsychiat., 2: 143-156, 1966.

116. Lewis, R.: Spontaneous autoimmune diseases of domestic animals. In International Review of Experimental Pathology, G. W. Richter and M. A. Epstein, (Eds). Academic Press, Vol.13, pp. 55-82, 1974.

117. Mandel, L.R., Ahn, H.S., VandenHeuvel, W.J., and Walker, R.W. Indoleamine-N-methyl transferase in human lung. Biochem. Pharmacol. 21: 1197-1200, 1972.

118. Marjerrison, G., Krause, A.E., and Keogh, R.P.: Variability of the EEG in schizophrenia: Quantitative analysis with a modulus voltage integrator. EEG and Clin. Neurophysiol., 24: 35-41, 1967.

119. Matthysse, S. and Haber, S.: Animal Models of Schizophrenia. Unpublished manuscript, 1974.

120. McGhie, A. and Chapman, J.: Disorders of attention and perception in early schizophrenia. Brit. J. Med. Psych., 34: 103-116, 1961.

121. McNeil, T.F. and Kaij, L.: Obstetrical notations of mental or behavioral disturbance. J. Psychosom. Med., 17: 175-188, 1973.

122. Mednick, S.A.: Breakdown in individuals at high risk for schizophrenia. Possible perinatal factors. Mental Hygiene, 54: 50-63, 1970.

123. Mednick, S.A., Mura, E., Schulsinger, F., and Mednick, B.: Perinatal conditions and infant development in children with schizophrenic parents. Social Biol., 18: 103-113, 1971.

124. Mednick, S. and Schulsinger, F.: Some premorbid characteristics related to breakdown in children with schizophrenic mothers. In The Transmission of Schizophrenia., Rosethal, D. and Kety, S. (Eds)., Pergamon Press, Oxford, p. 267, 1968.

125. Mednick, S. and Schulsinger, F.: In Children at risk: the search for the antecedents of schizophrenia. Part 2, Schizophrenia Bulletin, 9: 1974.

126. Moore, M., Nathan, D., Elliot, A.R., and Laubach, C.: Encephalographic studies in schizophrenia (dementia praecox). Am. J. Psychiat., 89: 801-810, 1933.

127. Moore, M., Natan, D., Elliot, A.R., and Laubach, C.: Encephalographic studies in mental disease. Am. J. Psychiat., 92: 43-67, 1935.

128. Morel, B.A.: Traite des maladies mentales. V. Masson, Paris, pp. 565-566, 1860.

129. Moss, R.L., and McCann, S.M.: Induction of mating behavior in rats by luteinizing hormone-releasing factor. Science, 181: 177-179, 1973.

130. Mudd, S.H. and Freeman, J.M.: $N^{5,10}$-methylenetetrahydrofolate reductase deficiency and schizophrenia: a working hypothesis. J. Psychiat. Res. 11: 259, 1974.

131. Murphree, O.D., Dykman, R.A., and Peters, J.E.: Genetically determined abnormal behavior in dogs: Results of behavioral tests. Cond. Reflex, 2:3, 1967.

132. Murphy, D.L., Belmaker, R., and Wyatt, R.J.: Monoamine oxidase in schizophrenia. J. Psychiat. Res., 2: 221-247, 1974.

133. Murphy, D.L., and Donnelly, C.H.: Monoamine oxidase in man: Enzyme characteristics in platelets, plasma and other tissues. In Neuropsychopharmacology of Monoamines and their Regulatory Enzymes, Usdin, E. (Ed)., Raven Press, New York, pp. 71-85, 1974.

134. Murphy, D.L., and Weiss, R.: Reduced monoamine oxidase activity in blood platelets from bipolar depressed patients. Am. J. Psychiat., 128: 35-41, 1972.

135. Murphy, D.L., and Wyatt, R.J.: Reduced MAO activity in blood platelets from schizophrenic patients. Nature, 238: 225-226, 1972.

136. Murphy, D.L., and Wyatt, R.J.: Platelet monoamine oxidase in schizophrenics. Nature, 253: 659-660, 1975.

137. Nameche, G.F., and Ricks, D.F.: Life patterns of children who became adult schizophrenics. In Children at risk: the search for the antecedents of schizophrenia. Part 2. (N. Garmezy, Ed.) Schizophrenia Bulletin, 9: 1974.

138. Odegard, O.: The incidence of mental diseases in Norway during World War II. Acta Psych. et Neurol. Scand. XXIX, 4:333, 1954.

139. Odegard, O.: Hospitalized Psychoses in Norway: Time Trends 1926-1965. Social Psychiatry, 6:2, 53-58, 1971.

140. Oldstone, M.: Virus neutralization and virus induced immune complex disease. In Progress in Medical Virology, J. L. Melnick, Ed., Vol. 19, S. Karger, Basel, pp. 84-119, 1975.

141. Overall, V.E., and Gorham, D.B.: Brief Psychiatric Rating Scale. Psychological Reports 10: 799-812, 1962.

142. Pasamanich, B. and Knobloch, H.: Retrospective studies on the epidemiology of reproductive casualty: old and new. Merrill-Palmer Quarterly of Behavior and Development 12:7-26, 1966.

143. Pfaff, D.W.: Luteinizing hormone-releasing factor potentiates lordosis behavior in hypophysectomized ovariectomized female rats. Science, 182: 1148-1149, 1973.

144. Pick, A.: Uber porinare chronische Demenz (sog Dementia Praecox) im jugendlichen Alter. Prager medizinische Wochenschrift, 16: 312-315, 1891.

145. Pincus, J.H. and Tucker, G.: Behavioral Neurology. Oxford University Press, New York, 1974.

146. Pinel, P.H.: Traite Medico-Philosophique sur l'alienation mentale. Chez J. Ant. Brosson, Paris, p. 180, 1809.

147. Pollin, W., and Stabenau, J.: Biological, psychological and historical differences in a series of monozygotic twins discordant for schizophrenia. In Transmission of Schizophrenia, Rosenthal, D., and Kety, S. (Eds.) Pergamon Press, London, 1968.

148. Pritchard, J.C.: A Treatise on insanity and other disorders affecting the mind. Haswell, Harrington and Haswell, Philadelphia, p. 71, 1837.

149. Raskind, M.A., Orenstein, H., and Christopher, T.G.: Acute psychosis, increased water ingestion, and inappropriate antidiuretic hormone secretion. Am. J. Psychiatry 132, Sept. 1975.

150. Reisby, N.: Psychoses in children of schizophrenic mothers. Acta Psychiat. Scand., 43: 8-20, 1967.

151. Robins, L.N.: Deviant Children Grown Up. Williams & Wilkins Company, Baltimore, Md., 1966.

152. Rochford, J.M., Detre, T., Tucker, G.T., and Harrow, M.: Neuropsychological impairments in functional psychiatric disease. Arch. Gen. Psychiat., 22: 114-119, 1970.

153. Rosenberg, D.E., Isbell, H., Miner, E.J., and Logan, C.R.: The effects of N,N-dimethyltryptamine in human subjects tolerant to lysergic acid diethylamide. Psychopharmacol., 5: 217, 1964.

154. Rosenthal, D.: Some factors associated with concordance and discordance with respect to schizophrenia in monozygotic twins. J. Nerv. & Ment. Dis., 129: 1-10, 1959.

155. Rosenthal, R. and Bigelow, L.B.: Quantitative Brain Measurements in Chronic Schizophrenia. Brit. J. Psychiat. 121: 259-264, 1972.

156. Rosenthal, D., and S. Nagler: Neurological and physiological characteristics of the children of schizophrenic parents. In Children at risk: the search for the antecedents of schizophrenia. Part 2. (N. Garmezy, Ed.) Schizophrenia Bulletin, 9: 1974.

157. Rush, B.: Medical inquiries and observations upon the diseases of the mind. John Richardson, Philadelphia, pp. 259-260, 1818.

158. Saavedra, J.M., and Axelrod, J.: A specific and sensitive enzymatic assay for tryptamine in tissues. J. Pharmacol. Exp. Ther., 182: 363-369, 1972.

159. Sameroff, A.J., and Zax, M.: Perinatal characteristic of the offspring of schizophrenic women. In Children at risk: the search for the antecedents of schizophrenia. Part 2 (N.Garmezy, Ed.) Schizophrenia Bulletin, 9: 1974.

160. Schwartz, M., Aikens, A.M., and Wyatt, R.J.: Monoamine oxidase activity in brains from schizophrenic and mentally normal individuals. Psychopharmacol. 38: 319-328, 1974.

161. Schwartz, M.A., Wyatt, R.J., Yang, H., and Neff, N.: Multiple forms of monoamine oxidase in brains: A comparison of enzymatic activity in mentally normal and chronic schizophrenic individuals. Arch. Gen. Psychiat., 31: 557-560, 1974.

162. Segals, J.: La Paranoia. Arch. de Neurol. 13: 62-76, 1887.

163. Shagass, C.: Evoked brain potentials in psychiatry. Plenum Press, New York, 1972.

164. Shagass, C.: Personal communication, 1975.

165. Sherpherd, M.: Personal communication.

166. Shields, J., and Gottesman, I.I.: Cross-national diagnosis of schizophrenia in twins. Arch. Gen. Psychiat. 27: 725-730, 1972.

167. Simon, R.J., Fleiss, J.L., Gurland, B.J., Stellar, P.R., and Sharpe, L.: Depression and Schizophrenia in hospitalized black and white mental patients. Arch. Gen. Psychiat. 28 (4): 509-512, 1973.

168. Singal, D.P., Graf, P. and MacCrimmon, D.: Antibodies to Thymocytes in sera of normal controls and of patients with schizophrenia. J. Immunol. 114(4): 425-427, 1975.

169. Slater, E.: Psychotic and neurotic illness in twins. Med. Res. Council Special Report Series No: 278, London, 1953.

170. Slater, E., and Couie, V.A.: The Genetics of Mental Disorders, London, 1971.

171. Small, J.G., and Small, I.F.: Reevaluation of clinical EEG findings in schizophrenia. Dis. Nerv. Syst., 26: 345-349, 1965.

172. Sobel, D.E.: Infant mortality and malformations in children of schizophrenic women. Psych. Quart., 35: 60-65, 1961.

173. Spitzka, E.C.: Insanity. Its Classification, Diagnosis, and Treatment. Bermingham and Co., New York, pp. 149-152, 1883.

174. Stabeneau, J.R., and Pollin, W.: Early characteristics of monozygotic twins discordant for schizophrenia. Arch. Gen. Psychiat. 17: 723-734, 1967.

175. Stabeneau, J.R., Turpin, J., Werner, M., and Pollin, W.A.: A comparative study of families of schizophrenics, delinquents and normals. Psychiatry, 28: 45-59, 1965.

176. Subba Rao, P.V., Jegannathan, N.S., and Vaidyanathan, C.S.: Enzymic conversion of 3-hydroxyanthranilic acid into cinnabarinic acid by the nuclear fraction of rat liver. Biochem. J. 95: 628, 1965.

177. Sullivan, H.S.: Conceptions of modern psychiatry. William Alanson White Foundation, Washington, D.C. 1953.

178. Teuber, H.L.: Effects of focal brain lesions. In: Prospects for Research in Schizophrenia, Kety, S.S. and Matthysse, S., Eds., Neurosciences Research Program Bulletin, 10: 4, 381-384, 1972.

179. Thetford, P.: Panel Discussion on Research. In: The Schizophrenic Reactions. Cancro, R. (Ed.) Brunner/Mazel, New York, p. 273, 1970.

180. Tucker, G.T., Campion, E.W., Kelleher, P.A., and Silberfarb, P.M.: The relationship of subtle neurologic impairments to disturbances of thinking. Second Congress of the International College of Psychosomatic Medicine, Amsterdam, 1973.

181. Vartanian, M.: Personal communication.

182. Vogel, R.A.: Academicae praelectiones de cognoscendis et curandis praecipus corporis humanii affectibus. Vanderhoeck, Gottingen, 1772.

183. Walker, R.W., Ahn, H.S., Albers-Schonberg, G., Mandel, L.R., and Vandenheuvel, W.J.: Gas chromatographic-mass spectrometric isotope dilution assay for N,N-dimethyltryptamine in human plasma. Biochem. Med., 8: 105-113, 1973.

184. Welner, J. and Stromgren, E.: Clinical and genetic studies on benign schizophreniform psychoses based on follow up. Acta Psychiat. et Neurologica Scand. 33: 377, 1958.

185. Wender, P.H.: Adopted children and their families in the evaluation of nature-nurture interactions in the schizophrenic disorders. Ann. Rev. Med., 23: 355-372, 1972.

186. Wender, P.H., Rosenthal, D., Kety, S.S., Schulsinger, F., and Welner, J.: Cross fostering: a research strategy for clarifying the role of genetic and experiential factors in the etiology of schizophrenia. Arch. Gen. Psychiat. 30: 121, 1974.

187. Willis, T.: Two discourses concerning the soul of Brutes, which is that of the vital and sensitive of man. Thomas Dring, London, p. 193, 1683.

188. Willis, W.D., Ashkenazi, R., Willis, J.C., and Haber, B.: The role of monoamine oxidase in terminating the reflex effects of Paramethoxy-Phenylethylamine (PMPEA). Texas Reports on Biology and Medicine 31: 423-429, 1973.

189. Wilson, I.C., Lara, P.P., and Prange, A.J.: Thyrotropin-Releasing Hormone in Schizophrenia. Lancet, 2: 43-44, 1973.

190. Wise, C.D., Baden, M.H., and Stein, L.: Proceedings: Post-mortem measurements of enzymes in human brain: Evidence of a central noradrenergic deficit in schizophrenia. In: Enzymes in Neuropathology of Schizophrenia, Matthysse, S. and Kety, S.S. (Eds.) Pergamon Press, Oxford, 1974.

191. Witz, I.: Personal communication.

192. Woerner, M.G., Pollack, M., and Klein, D.F.: Pregnancy and birth complications in psychiatric patients: A comparison of schizophrenics and personality disorder patients with their siblings. Acta Psychiat. Scand., 49: 712-721, 1973.

193. World Health Organization. Report of the International Pilot Study of Schizophrenia, Volume 1. Geneva, 1973.

194. Wyatt, R.J., Gillin, J.C., Kaplan, J., Stillman, R., Mandel, L., Ahn, H.S., VandenHeuval, W.J. and Walker, R.W.: N.N-dimethyltryptamine - a possible relationship to schizophrenia? In Serotonin - New Vistas, Costa, E., Gessa, G. L., and Sandler, M. (Eds.), Raven Press, New York, pp. 299-315, 1974.

195. Wyatt, R. J., and Murphy, D.L.: Neurotransmitter-related enzymes in the major psychiatric disorder: II. MAO and DBH in schizophrenia. In: The Biology of the Major Psychoses: A Comparative Analysis., Freedman, D.X. (Ed.) Raven Press, New York, pp. 289-296, 1975.

196. Wyatt, R. J., and Murphy, D.L.: Low platelet monoamine oxidase activity and schizophrenia. Schizophrenia Bulletin, in press.

197. Wyatt, R.J., Murphy, D. L., Belmaker, R., Cohen, S., Donnelly, C., and Pollin, W.: Reduced monoamine oxidase activity in platelets: A possible genetic marker for vulnerability to schizophrenia. Science 173: 916-918, 1973.

198. Wyatt, R.J., Schwartz, M.A., Erdelyi, E., and Barchas, J.: Dopamine-beta-hydroxylase activity in the brains of chronic schizophrenic patients. Science, 187: 368-370, 1975.

199. Young, I.J., and Crampton, A.R.: Cerebrospinal fluid uric acid levels in cerebral atrophy occurring in psychiatric and neurologic patients. Biol. Psychiat. 8: 281-292, 1974.

200. Zubin, J. Personal communication, 1975.

SUBJECT INDEX

NAME INDEX
PERSONS NOT CITED IN BIBLIOGRAPHY

MIX
Papier aus verantwortungsvollen Quellen
Paper from responsible sources
FSC® C105338

If you have any concerns about our products,
you can contact us on
ProductSafety@springernature.com

In case Publisher is established outside the EU,
the EU authorized representative is:
Springer Nature Customer Service Center GmbH
Europaplatz 3, 69115 Heidelberg, Germany

Printed by Libri Plureos GmbH
in Hamburg, Germany